ESSENTIAL

MERCEDES SL

190SL & PAGODA MODELS

ESSENTIAL
MERCEDES SL
190SL & PAGODA MODELS

THE CARS AND THEIR STORY 1955-71

LAURENCE MEREDITH

Published 1997 by Bay View Books Ltd
The Red House, 25-26 Bridgeland Street,
Bideford, Devon EX39 2PZ, UK

Edited by Mark Hughes
Typesetting and design by Chris Fayers & Sarah Ward

ISBN 1 870979 89 3
Printed in Hong Kong

CONTENTS

A PROUD HERITAGE
6

THE RANGE IN BRIEF
14

THE 190SL
24

THE PAGODAS
42

THE CLASSIC SLs TODAY
72

APPENDIX
77

A Proud Heritage

It is hardly surprising that the company which invented the motor car – albeit in a primitive form – has continued to create some of the world's most coveted automobiles. For more than a century the famous three-pointed star has been a recognised symbol of excellence the world over. It represents everything for which the Mercedes-Benz marque stands – chiefly unrivalled standards of engineering integrity and build quality.

Yet there is always more. Leaders in the field of innovative technology and both primary and secondary safety, the Unterturkheim-based parent company, Daimler-Benz, has fostered a laudable policy of employing some of the most talented and gifted engineers throughout its long and sometimes turbulent history. Such is the economic standing of this respected company in Germany that the old adage, 'when Daimler-Benz sneezes, the rest of the country catches a cold', is by no means an exaggeration.

After the glorious days of the 1930s, when the famous state-backed 'Silver Arrows' Grand Prix cars displaced and

One survivor from the 1952 300SL sports racing coupé line that inspired the classic families of SL road cars of the 1950s and '60s.

comprehensively conquered traditional race winners from the stables of Bugatti and Alfa Romeo, the Second World War saw the fortunes of Daimler-Benz change dramatically. Allied bombing raids led to the destruction of factories and, after the cessation of hostilities in 1945, the long, arduous process of rebuilding began.

What remained intact, however, was the spirit to continue. Key people such as engineering director Fritz Nallinger and his assistant Rudi Uhlenhaut, who would go on to become chief development engineer, had the same enthusiasm for the company, and more particularly for its sporting activities, that they had so demonstrably portrayed some 10 years earlier.

Until the German currency reform in 1948, Daimler-Benz, along with Volkswagen and the fledgling Porsche concern, struggled to get by. The senior of the two

Through 1952 the racing 300SL 'Gullwing' coupés were all-conquering, with victory in the arduous five-day Carrera PanAmericana perhaps their most convincing success. This is the winning car of Karl Kling and Hans Klenk pausing while its smashed windscreen, caused by an impact with a buzzard, is examined during the early stages of this event through Mexico.

Stuttgart manufacturers produced around 200 cars by the end of 1946 and 1000 by the end of the following year. Some 5000 cars rolled off the Sindelfingen assembly lines in 1948, but once the international authorities controlling Germany permitted her companies to resume trade in export markets, output from the German, Polish and Turkish workforce spiralled to 34,000 cars in 1950.

Initially, profits were modest but encouraging and, with Uhlenhaut's distinguished background both as a designer and an accomplished test driver, it was only a matter of time before motor sport figured high on the agenda at Daimler-Benz board meetings – the name

Mercedes–Benz needed to be re-established as a top race winner. This would not only bring prestige to a company that craved this but, because of the publicity generated in the international motoring press, it was also a good deal cheaper than advertising.

A small delegation headed by the great Alfred Neubauer, who, according to legend, did an excellent and irreverent impression of Adolf Hitler, was despatched to watch the 1951 Le Mans 24 Hours, which was won by Peter Walker and Peter Whitehead in a C-type Jaguar. Armed with a case of technical data gathered from this race about the opposition and their strategies, Neubauer returned to Stuttgart.

As a result, Nallinger and Uhlenhaut lost no time in designing a new racing car to compete at Le Mans the following year. Grand Prix racing was out of the question at this stage due to a lack of funds, but Le Mans would suffice for starters. The technical team had done its homework well, because the new 3 litre six cylinder 300SL Coupé scored an outright victory at the Sarthe

Three views of the sensational 300SL racer of 1952. Initially the doors were totally impractical (above), the full gullwing design yet to appear. Possibly the greatest driver of the pre-war era, Rudi Caracciola (right), made a comeback in the 300SL that season, and is seen at Bern, Switzerland, shortly before the crash that ended his career. The 300SL also raced in roadster form (below), two of these versions pictured at the Nürburgring with eventual winner Hermann Lang trailing Karl Kling.

Seen with the coupé version of the eight-cylinder 300SLR, engineering genius Rudi Uhlenhaut was largely **responsible for the first-generation and second-generation SL road cars. The row of parked cars includes a 190SL.**

circuit in 1952, and a similar car finished second to rub salt in the wounds. It was a remarkable achievement – the Silver Arrows were back. Further outright victories were scored in the same year in the Mille Miglia and Carrera PanAmericana races – a most satisfying cake-icing exercise for the Daimler-Benz board.

Then, much to the irritation of Enzo Ferrari in particular, Mercedes withdrew from sports car racing at the end of the season to concentrate upon, among other projects, the design of a new Grand Prix car. However, the 300SL racers of 1952 had made a big impression on all those who witnessed their domination of sports car racing. Their mark was so indelibly printed on the imagination of the Austrian-born, New York-based sports car importer and enthusiast Max Hoffmann that, according to popular but officially unconfirmed lore, he

put in a request for no fewer than 1000 road-going versions of the 300SL.

Daimler-Benz was only too glad to oblige him; America in particular was a large, rich country populated by people who were hungry to part with dollars for expensive European sports cars. Mercedes also needed dollars. Several changes were made to the 300SL road car between prototype and production and, because its existence was announced well in advance of its official public launch at the New York Motor Show in February 1954, the motoring world was expecting it. What they did not expect, however, was the simultaneous debut of the 300SL Coupé's smaller sister – the 190SL.

The 300SL 'Gullwing', as it was dubbed because of its unique top-hinged doors, caused an audible sensation among those present when the wraps were taken away on that day in 1954. To use a much-abused word, this was the first true 'supercar' of the post-war period. Its gorgeous two-seater bodywork was aerodynamically sound. Its six-cylinder 3-litre engine – canted over at an angle to allow for a low bonnet line – was race-derived and capable of propelling the car to a top speed of

As the fastest road car of its time, the 300SL Gullwing mixed with the rich and famous – this is film actress Sophia Loren.

155mph (250kph) with the optional high final drive ratio, and also boasted direct fuel injection, the world's first production car fitted with this system.

However, the Gullwing had several drawbacks. At $6820 it was wickedly expensive and clearly aimed at the mega-rich. As a serious racing car for the road it was a riot, but the high-pivot, swing-axle rear suspension gave tricky on-the-limit roadholding in inexperienced hands, the doors could not be opened in the unhappy event of inverting the beast, and the broad, high sills were hardly conducive to dignified entry and exit by women in short skirts – by this time women figured significantly among Daimler-Benz's customers.

In 1957 the problems of the 300SL Coupé were addressed with the launch of the 300SL Roadster which, thanks to a heavier spaceframe chassis and 'rag-top' aerodynamics, was some 20mph (32kph) slower than the car it replaced. But, by way of compensation, it had conventional, front-hinged doors, a low-pivot swing axle for improved roadholding, and was altogether more 'user-friendly'. While everyone stood in awe of the Gullwing, it

The two versions of the 300SL in production form – Mercedes built 1400 examples of the Gullwing (1954-57) and 1858 of the Roadster (1957-63).

Styling study for the 190SL: of unknown scale, this model looks very similar to the finished car, but note the absence of any representation of windscreen wipers and the slightly more squared-off upper corners of the grille.

was the Roadster version that motoring enthusiasts wanted to drive. Meanwhile the 190SL – the 'baby' sporting Merc – soldiered on in production, for it was not only popular and in heavy demand, but also extremely profitable for Daimler-Benz.

It was not until the early 1960s that the need to replace it became inevitable. By this time the automotive world had changed. Brutally fast sports cars which owed their appeal to straight-line performance, without much consideration being given to roadholding capabilities, gave way to a new era in which the radial tyre would revolutionise chassis engineering. Uhlenhaut began with a fresh sheet of paper for the second-generation SL. In launching the relatively tame 230SL in 1963, Daimler-Benz made a conscious effort to steer the company and its products in a different direction. The company left the pursuit of ultimate speed to others and, with the wisdom of hindsight, it was correct to do so.

With the notable exception of Porsche, virtually all the other 'supercar' manufacturers struggled to find customers through that decade. In Italy, Ferrari was eventually swallowed by Fiat, Bizzarrini ceased trading altogether, Lamborghini stumbled from crisis to crisis, and Maserati was taken over by Citroën. In Britain, Jaguar was absorbed into the abortive British Leyland conglomerate, while for the Swiss manufacturer of exclusive grand tourers, Peter Monteverdi, the end came in 1976.

Daimler-Benz succeeded where others failed. The company's marketing people carried out extensive research and did their homework thoroughly. Daimler-

Benz knew that there would always be a large number of sporting enthusiasts who were sufficiently well-heeled to afford a Mercedes, but also realised, as a result of market research, that the vast majority of its customers could not afford to buy a new one until they were over 40 years of age, by which time few would tolerate the many drawbacks – excessive engine noise, lack of reliability, exorbitant running costs, insufficient luggage space and conspicuously styled bodywork – of contemporary exotica from Italy.

What the Stuttgart company provided, therefore, was a comfortable, quiet, safe and peerlessly engineered package with restrained, conservative styling and an engine which, although satisfyingly powerful, was rarely going to raise the heart rate to a potentially dangerous level. Some of Daimler-Benz's directors fitted this category exactly, which is why they had such a fundamentally good understanding of customers' needs.

For the typical Italian driving enthusiast the removal of the gear stick is akin to castration, but for Daimler-Benz customers the availability of an automatic gearbox in the 'Pagoda' range provided welcome relief from the wearying need to change gear. A slim, tight-fitting bucket seat – so essential in a Ferrari – was an unnecessary and uncomfortable accoutrement better suited to a fit, young thing in the opinion of the slightly portly Mercedes owner. In contrast to the elegant, wood-rimmed steering wheel fitted to the majority of Italian sports cars, Mercedes drivers always sat behind a thick-rimmed wheel of truck-like proportions. The physical effort involved in driving was largely minimised in an effort to reduce stress suffered after a day at the office. And, whereas pukka sports cars traditionally had a plethora of dashboard instruments, Daimler-Benz was resolute in its long-held policy of providing as few as possible in the interests of safety and reduced levels of driver fatigue – a philosophy which remains intact to this day.

Styling study for the 230SL: dwarfed by a real car, this model has intricate wing vents that were not adopted for production.

Dependable and over-engineered in characteristic fashion, both the 190SL and the Pagodas have stood the test of time well and a high percentage of both types survive today. In its styling, fixtures and fittings, the 190SL is so obviously a car that belongs to the 1950s – part of its appeal – but many argue that the shape of the 230SL, 250SL and 280SL has never dated. There is no doubting the Pagoda's classic status, accorded for many more reasons than the prestige of the large star in the centre of its wide radiator grille.

The ability to cover 'starship' mileages without major mechanical surgery allows these cars to last just about forever if maintained properly. They resist corrosion rather well and boast almost unrivalled levels of comfort. A Pagoda is also fun to drive, possessing the handling and roadholding properties of a well-controlled Scalextric car. But because it was intended as and perceived to be the 'sensible brogue shoes' of the 1960s sports car world and is clearly without the charisma and dramatic styling of a Ferrari – the Pagoda is one of the few worthwhile classics that has remained affordable, even throughout the heady days of the late 1980s.

Owning one of the classic SLs, first or second generation, has always meant more to Mercedes *aficionados* than the mere and temporary custodianship of a well-assembled collection of steel, alloy, rubber and plastic componentry – it is more a way of life. Buy an SL and you automatically become a member of a world-wide brotherhood of enthusiasts who appreciate the special aura that these cars exude.

As we approach the end of the 20th Century, running one as an everyday road car can still make sense. Each model is well catered for when it comes to replacing broken or worn parts, and there are several clubs world-wide whose more knowledgeable members are always prepared to help in locating the rarer cars. Seemingly, SL ownership has but one drawback: buy one and enjoy it for what it is, and you will never be able to let go of it or change it for anything else, no matter how strong the appeal of other marques may become from time to time.

While I was researching this book, a prominent British member of the Mercedes-Benz Club and a collector of classic Mercedes told me that he had sold his brand-new 500SL Mercedes after just six months because it was "too good, too perfect", and performed its essential role without requiring much in the way of driver input. In short, it had none of the desirable and inherent personality of the SLs of the 1950s and '60s. It is the character, above all, of the classic SL that continues to command the respect of collectors and car enthusiasts all round the world.

THE RANGE IN BRIEF

After the war the production of traditional sports cars fell predominantly into the domain of British manufacturers. MG, Austin-Healey, Jaguar, Triumph and Morgan, to name but a few, all made open two-seaters that, apart from the Jaguar XK120, were simple in construction and often crude in their appointments. They were, however, eminently affordable and eligible for competition work, at which they excelled, and they enjoyed enormous success in the US.

Until the advent of the Porsche 356 coupé and cabriolet in 1950, Germany was largely without a proper sports machine of its own. Through the 1950s BMW concentrated on making motorcycles and luxury saloons, while the pretty Volkswagen Karmann-Ghia launched in 1955 could hardly be deemed a sports car, with its 1192cc 30bhp Beetle engine, despite aspirations promoted by its two-seater bodywork. The cost of a 300SL Gullwing meant that owning one was out of the question for the majority of sports car addicts, but the 190SL – a masterpiece of marketing and design – was just in reach of the comfortably off, and was to make significant inroads into the British stranglehold on sports car sales in North America.

Sitting on the same length wheelbase as the Gullwing, the 190SL's styling was altered from the pre-production prototypes to ape the Gullwing's elegant lines more closely, particularly around the nose and radiator grille – more than 40 years on it is still possible to confuse the two models from a distance. Whereas the Gullwing was a highly complex car, the 190SL was just the opposite, easily maintained and serviced without the need for special tools and engineering knowledge.

The 190SL's looks alone caused quite a stir when it was originally announced, its curvaceous wings and

Inviting, comfortable and uncluttered as this artist's impression portrays the interior to be, the cabin is even more enticing in reality. As with future SLs, the 'baby' 190SL was also available with a smart hard-top.

US advert from the May 1956 issue of *Road & Track*, the leading car magazine in the 190SL's most important export market.

top could be fitted. With the soft-top down, this gave wind-in-the-hair motoring enthusiasts what they were looking for; the 190SL felt like a proper sports car, even if it did not exactly perform like one.

Because the 190SL had to be profitable and sell in volume, it was based on the W120 – or 180 saloon – and therefore drew heavily on the Sindelfingen parts bin. In contrast to the Gullwing's multi-tubular spaceframe chassis, little sister's steel bodywork (with aluminium doors, bonnet and boot lid) was of unitary construction, like the 180 saloon – a much simpler and cheaper method of manufacture. The loss of torsional rigidity inherent in all convertible bodies was largely compensated for in the 190SL by the longitudinal, box-section side members and central tunnel of the chassis, which made for a strong structure that gave little scuttle-shake even on the roughest road surface.

Although the 190SL's floorpan was 25cm (10in) shorter than the 180 saloon's, the latter's suspension system was bolted onto the two-seater without major modification. At the front there were conventional unequal-length wishbones, coil springs and telescopic shock absorbers. Rear suspension was by coil springs, trailing arms, swinging half axles and telescopic dampers.

wheelarch eyebrows giving it a most purposeful appearance. In period fashion the bodywork was tastefully trimmed – but not excessively so – with chrome-plated or alloy mouldings. Although its 1.9-litre four-cylinder engine was not comparable in performance to out-and-out British sporting machinery, it differed significantly from the 300SL coupé in having a convertible top. From the beginning the 190SL came as a coupé with a detachable hard-top or as a convertible to which a hard-

Safety and security were, for the time, unusual aspects to emphasise – this 1962 advert is from *Réalités*, a French 'lifestyle' magazine.

The system was fully independent, but not beyond criticism by those who failed to cope with the supposed vagaries – sudden tail-end breakaway and rear-wheel 'tuck-in' – of the rear swing axles. At a time when Porsches, Beetles and the Gullwing all sat on 15in diameter road wheels, the 190SL was unusually modern in having 13in wheels, which not only helped to reduce unsprung weight but usefully lowered the car's centre of gravity – immensely beneficial when it came to negotiating high-speed corners.

Daimler-Benz's four-speed all-synchromesh gearbox and famous recirculating ball steering box were fitted, but the hydraulically operated drum brakes at all four corners differed markedly from the 180 saloon's and more closely resembled the Gullwing's. Supplied by ATE, the inner and outer castings of the drums were made from cast iron (the 300SL's were alloy) and had impressive-looking external cooling ribs. Braking, of course, was one area of engineering development where the Germans lagged behind the British. Working in conjunction with Jaguar in particular, Dunlop had demonstrably proved – principally at Le Mans – that disc brakes were superior. Daimler-Benz would eventually adopt discs for the 300SL Roadster, but not until 1961, two years before production of the 190SL ceased.

C'est uniquement en se sentant en sécurité – donc à l'aise et détendu – que l'on peut apprécier pleinement les avantages d'une automobile. C'est pourquoi chaque Mercedes-Benz est l'objet, depuis la création jusqu'à la production, de nombreux efforts efficaces qui lui confèrent la plus grande sécurité. De grandes qualités routières, une construction solide, une finition de haute qualité et beaucoup d'accessoires – de la serrure à double cran au pare-soleil capitonné – donnent cette impression, particulière à une Mercedes-Benz: une sécurité tangible.

When *Road & Track* tested the 190SL in October 1955 it commented: 'The outstanding achievement of the 190SL is without a doubt its quality in design and workmanship. But a close second is the general feeling of solidity which it immediately conveys… Obviously the 190SL is not a "bomb" in acceleration; nor is it "Super Leicht", but the acceleration times are very good for a 2-litre car of this weight. It is easy to overlook the fact that this engine is very small by US standards, for 1897cc is only 115.7cu in.'

The *Super Leicht* (Super Light) appellation was something of an enigma – the product of the marketing department aiming to identify the car more closely with the Gullwing – but few customers complained about this minor misnomer.

It is only natural that the 300SL Coupé and the 190SL were compared with each other by contemporary journalists, although the two cars are as different as wine and whisky in many respects, and it is most surprising that a number of seasoned writers actually preferred the less powerful model.

The author of a test in the April 1957 issue of *Sports Car World,* for example, commented: 'After living with a 300SL, I made the mistake of assuming that the "lesser" car would be a letdown. I was wrong. The 190SL is just as exciting, in a quieter, more subtle way, as the 300SL, and it's my feeling that for most mortals it's actually a more desirable car.

'In the 300SL you're over-gunned for the road. In the 190SL you're armed just right. It corners more securely than the 300SL, it has the same excellent steering, a similar full-synchro gearbox, the same quality finish throughout and better rear suspension. Its beauty of line and many of its dimensions approach those of the 300SL.

'But it's a car that you might not mind turning your wife loose with – something not many 300SL owners are doing, you can bet. For a sports touring car – not a competition car – the 190SL is about as close an approach to perfection as any of us are likely to see, and for the kind of connoisseur's car it is, it's not expensive. But if you want a car for winning class E races, keep looking; this is not the machine.'

A small number of owners, particularly in the US, did use their 190SLs for club racing and had a lot of fun with them. For this purpose Daimler-Benz provided a small Perspex aeroscreen for the driver's side to replace the standard windscreen, and lighter aluminium doors with 'arm-rest' cut-outs. Most drivers removed the bumpers for competition work to save a little weight, but they still did not catch the Jaguar XKs.

Journalists typically found that the car had a top speed of between 103–105mph (166–169kph), the benchmark figure of 60mph arriving from rest in an impressive 11sec. These figures compare favourably with the 1600cc pushrod-engined Porsche 356, which *The Autocar* discovered in its 1956 road test would accelerate from 0–60mph in 15.3sec and reach a maximum speed of 102mph (164kph). It is clearly paradoxical, therefore, that many writers considered that the Porsche was a 'proper'

sports car while the 190SL was not. The author of *The Autocar's* Porsche test comments: 'In its present form the Porsche behaves more like an orthodox high-performance sports car... Many high performance cars are often seen to be driven quite slowly, but observation of many Porsches in their native land reveals that these cars are bought by people who like to drive fast.'

No-one would doubt the genuine sporting nature and prowess of a 356 Porsche but, as *The Autocar's* 190SL report of 1958 acknowledges, 'Several continental journalists who tested the 190SL soon after it was launched considered it to be rather sluggish and lacking in the performance expected from such a car...' Interestingly, Daimler-Benz appeared to encourage this non-sporting image by describing the 190SL as a 'sports tourer' which, when fitted with the small aeroscreen and cut-down doors, made it possible to 'participate with success in sporting events on a modest scale.'

Typically, the independent-minded individuals who bought the 190SL did not care about the blurred definition of a sports car, or about Daimler-Benz's description, and voted with their pockets. In the first year's production 1717 190SLs were sold, but this figure more than doubled to 4032 the following year. Annual sales figures then averaged roughly 3000 until the end of production early in 1963.

In contrast to British manufacturers, who considered it necessary to launch wholly new models at regular intervals, the Germans tended to improve and refine the same model over a long period of time, and in this respect Daimler-Benz was no exception. By 1963 the 190SL was a better car than it was at launch nine years earlier owing to many year-on-year improvements, but these were very much in the nature of under-the-skin modifications, the car remaining outwardly unchanged virtually throughout.

When the 190SL finally bowed out, it did so to make way for a genuinely superior motor car. The 230SL made its debut at the Geneva Motor Show in March 1963, two years after the launch of the Jaguar E-type. Although instantly recognisable as a Mercedes-Benz, it was a very different animal from its sporting predecessors. Again, there were problems in some quarters as to its role: it clearly was not a sports car in the accepted sense, it certainly could not compete on performance with the E-type (yet it cost considerably more), and it definitely was not 'Super Light' as its badge suggested.

What journalists and Daimler-Benz's loyal clientele came quickly to realise, though, was that the Pagoda was the clever product of exceptionally clever people. Based on the 220SE saloon, it was smartly styled and presented

Brochure illustration had moved to photography by the time the 230SL appeared. Versatility is stressed here – the car's luggage capacity and roof choices.

Brightly executed and fluent, Bracq's lines are timeless and remain unforgettable. Gone were the rounded curves of the previous generation SLs, to be replaced by 'razor' edges to reflect more accurately the sobriety and respectability of the large cross-section of customers who queued up to buy Pagodas. With its gaping radiator grille and wide track, the 230SL appears to have a short wheelbase in relation to its overall breadth, a clever illusion that gives the large body the looks of a compact and much smaller car than it actually is.

The structure of the car, and more particularly the hard-top, made the 230SL one of the safest cars in the world – if not *the* safest – at this time. This was largely down to the work of Béla Barényi, manager of the pre-development department (closed in 1974), a foresighted genius whose inventions – safety steering column, side-impact bars, rigid passenger cell and impact-absorbing crumple zones to give just a few examples – have saved the lives of hundreds of thousands of people throughout the world.

Today, the results of Barényi's research and experimentation, particularly during the 1950s, are recognised by the motor industry as being fundamentally important, if not essential, to automotive design. Barényi loved motor cars and enjoyed driving them even more. Like Uhlenhaut, he had an intuitive feel for their inherent faults, and devoted much of his life to developing means and ways of correcting them.

The Daimler-Benz patent (DPB 1.069.008) for the Pagoda roof – so strong that it can accept loads of more than 1000kg (2205lb) without deforming – was finalised as early as 1956. However, the 230SL (or W113 in Daimler-Benz's complex numbering system) was not the first Mercedes with such a roof. Barényi carried out exhaustive tests with a similar structure fitted to the

the right image. Under the bonnet was a new 2.3-litre six-cylinder engine driving the rear wheels through a choice of manual or automatic transmission. The top speed of 115mph (185kph) – 34mph (55kph) less than the E-type – was deemed quite sufficient for prevailing road conditions in 1963, and it achieved this effortlessly. Its relatively pedestrian straight-line performance, however, demonstrated what a deceptive piece of machinery this was, as Rudi Uhlenhaut proved on a French race track, the little Montroux circuit close to Geneva, when he put up a comparable lap time to a Ferrari 250GT driven by Mike Parkes.

From the beginning of production this was a two-seater convertible with a detachable hard-top offered as an extra-cost option. Apart from the roof and laudably large window glass, the 230SL's most distinguishing features were the vertically-stacked headlamp clusters enclosed by large one-piece lenses – except on American-specification cars which had circular headlamps – that had been seen first on the 300SL Roadster and remained a feature of several Mercedes models well into the 1970s.

Conservatively styled by Paul Bracq, the graceful bodywork was not dramatic in the way of a Ferrari or Maserati, but purposeful and attractive just the same.

There is a lack of attention to detail, surprising for Mercedes-Benz, in this 1964 *Road & Track* advert: all 230SLs sold in the US were fitted with sealed-beam headlamp assemblies, which look very different. Evoking the romance and freedom of touring (facing page) – this striking image is from the original 230SL sales brochure.

motoring – and so the top of the roof was divided into three distinct parts.

The centre section was concave – not the result of an elephant sitting on it as a popular contemporary cartoon depicted – to achieve the 'low' stance, and the two outer sections were raised. This not only increased headroom but also allowed a large window glass area. To have used a uniform, flat roof panel would undoubtedly have made the 230SL appear gawky and top heavy. What Barényi achieved, therefore, was a clever three-way compromise between aesthetics, strength and safety.

As is usually the case when compromises have to be made, though, there was just one disadvantage to this

W111, an experimental car based on the 220S saloon that did not go into production.

A low roof line in a sporting car, of course, was desirable from an aesthetic point of view but, according to Barényi, this could only be achieved at the expense of a relatively small window glass area. To have taken the latter route would have been unacceptable on safety grounds – good all-round visibility is essential for high-speed

design – it was less than aerodynamically sound. Barényi was well aware of this, but considered it to be relatively unimportant bearing in mind the nature of the car. Surprisingly, perhaps, the soft-top gave a slightly improved drag co-efficient, and a largely insignificant higher top speed of 2mph (4kph). As the majority of drivers would never reach anything approaching top speed, this did not matter – safety was much more important.

If Barényi's efforts in this area were pioneering, and they most certainly were, Uhlenhaut broke new ground in his laudable initiative in ridding the world of the curse of crossply tyres. These quite dreadful things had given the Gullwing an appalling and undeserved reputation as an evil high-speed handler, a trait that was blamed by the ignorant on the swing-axle rear suspension, and Uhlenhaut recognised that it was time for the tyre industry to pull up its socks.

As the principal responsible for the 300SL Coupé, Uhlenhaut was quietly disgruntled by the considerable criticism that had been squarely aimed at 'his' masterpiece. This adversely reflected on his personal choice – borne out by the deficiencies of the popular alternatives he might have considered – of the swing-axle system. In my opinion, crossply tyres proved to be the bane of most 1950s and '60s sports cars, but the 300SL suffered in particular because it was so very much faster than the others. Uhlenhaut, never a man to mince his words, forcefully pressed Continental and Firestone Phoenix to produce a new tyre – the radial – that could cope with the demands of his chassis.

Michelin X radials had been fitted to the 300SL previously, but they were found to be unsuitable and these cars even had a sticker on the dashboard warning drivers not to exceed 125mph. However, the Continentals fitted to the 230SL proved to be just about perfect (although one or two journalists thought otherwise), and one reason why these cars still feel modern and safe to drive today. Curiously for a car in this class, and in contrast to contemporary Italian exotica, the Pagodas were all fitted with plain steel wheels; alloys were offered on 280SLs from late 1969, but only as extra-cost options. Many thought the alloys to be so attractive that they fitted them retrospectively to earlier cars.

With the considerable advantages of fuel injection, the straight-six single overhead camshaft engine was an almost turbine-smooth jewel but, notwithstanding the car's overall ability to run rings around ostensibly much lighter, faster cars on the road, there was a perception that the 230SL needed more power. Daimler-Benz answered the critics by announcing a revised car, the 250SL, in

December 1966, its public debut occurring three months later at the Geneva Motor Show.

The 250SL was outwardly similar, but its new engine was based on that of the 250 saloon. Capacity was increased to 2496cc (152.3cu in) by lengthening the stroke by 6.2mm (0.24in) to 79mm (3.11in), but the bore was left unchanged at 82mm (3.23in). The cylinder head was reworked and had larger ports and valves, while a stronger seven-bearing crankshaft took the place of the previous four-bearing version. Although power output at 150bhp (DIN) was no greater than that of the engine it replaced, torque increased to 144.5lb ft at 4200rpm.

With a slightly higher final drive ratio, the 250SL had a marginally higher top speed both in four-speed manual and automatic gearbox forms. A five-speed manual gearbox was also offered as an extra-cost option on the 250SL, but this was rarely requested because, by this time, the majority of customers had so tired of using a gear lever that automatic gearboxes had become overwhelmingly more popular.

The 250SL was short-lived and just 4752 examples (compared with 19,831 230SLs) were built before the model was replaced by the 280SL just a year later. Again, it was basically the same car, but with a larger engine and a number of under-the-skin improvements. The engine was to the same basic seven-bearing design as the 250SL, with the stroke left unaltered but the bore increased to 86.5mm (3.41in) to give an overall size of 2778cc (169.5cu in). Power output went up accordingly to 170bhp (DIN) at 5700rpm and torque to 177lb ft at 4500rpm, but the advantages of this larger motor were soon negated to some extent on American-specification cars as these were fitted with power-sapping exhaust emissions equipment, output being quoted at 180bhp (SAE) – 15bhp less than European counterparts according to the more generous SAE rating system.

It was during the 1960s that American consumer groups concerned about road safety began to exert considerable pressure on motor manufacturers to change the way in which they constructed motor cars. Led by a lawyer, Ralph Nader, these groups had considerable influence, many of their aims eventually becoming enshrined in legislation.

Largely thanks to Barényi's efforts, Daimler-Benz was well ahead of its competitors in the fields of primary and secondary safety, but, to accommodate the requirements of increasingly stringent federal laws, the 280SL was a necessarily different animal from its two predecessors. The later cars were more refined in many ways but they were also considerably heavier. Since the 280SL also had

In 1971 the Pagoda was replaced by an altogether more sophisticated animal – the third-generation SL – that reflected growing concern over road safety and exhaust emissions. These were fine cars, seen here in 350SL and long-wheelbase 450SLC forms, but aficionados mourned the 280SL's passing.

rubber-bushed suspension to cut down on the need for the regular greasing, a servicing nuisance on the 230SL and 250SL, the handling became less crisp.

This 'softness', of course, found favour in North America, but proved slightly less popular with German customers who traditionally preferred firm suspension. Some purists may have favoured the original 230SL for its relative simplicity, but the appeal of the extra power of the larger-engined cars is reflected in sales figures. Exactly 23,885 280SLs found customers between 1967 and the end of production in 1971. Interestingly, 12,927 280SLs were exported to North America – more than half of the total production of this version.

These figures are impressive for such a relatively expensive motor car, although the affluent era in which the Pagodas were made certainly helped their success. However, as technically advanced and superb to drive as they were, the low-pivot swing-axle suspension came in for increasing criticism towards the end of the 1960s.

As journalists began to discover that other manufacturers had made advances in chassis technology, it became obvious that the Pagoda's days were numbered. Towards the end of the decade, Daimler-Benz was well aware that a replacement would be needed to take the company successfully into the 1970s, despite the general view that the 280SL was just about peerless.

In 1971 the classic Pagoda was succeeded by the 350SL, a wholly new model that marked the beginning of the third-generation SLs. The new car boasted state-of-the-art chassis technology that included anti-dive suspension geometry, a much larger and stronger body and an engine that for the first time since the Gullwing had 'real' power – up to 230bhp in 4.5-litre form – commensurate with grand touring aspirations.

Many mourned the 280SL's passing and understandably so. In many ways its successor was a far superior animal, if lacking in character, but the legacy of the 1960s makes the Pagodas the classic Mercedes SLs.

THE 190SL

The car in this period publicity shot has whitewall tyres, accoutrements that were more popular in North America than in Europe, and optional front bumper overriders.

With its increasingly complex *autobahn* network, Germany was ideal for enjoying the advantages of a sports tourer during the 1950s. Long tracts of beautifully built motorways were largely free of the congestion we all experience in Europe and America today, and, by driving the 190SL at its maximum speed of 105mph (169kph), 'commuting' between one city and another was not only fun and exciting but also took very little time. Good aerodynamics also suited the 190SL to long-distance work, and aided its impressive fuel consumption of up to and occasionally above 30mpg in favourable conditions, even at high speeds.

When it went into production, the 190SL was a little heavier than Daimler-Benz had originally planned, but everything about it reeked of quality. Although body rot would eventually prove to be its principal enemy – 190SLs can corrode with a vengeance if left unchecked – every component looked and felt as if hewn from solid.

Panel fit was impressive and the paint finish flawless, but this was no less than was expected from Daimler-Benz. As one writer commented in *Road & Track*'s October 1955 test: 'The outstanding achievement of the 190SL is without doubt its quality in design and workmanship. But a close second is the general feeling of solidity which it immediately conveys. It weighs exactly 2500lb with a full tank, and it feels like a 4000lb car on the road.'

Although made to the highest possible standards with a sturdy frame and woven mohair, the soft-top had a plastic rear window at a time when the Karmann-built Volkswagen Cabriolet had one made of glass. The heavy double-skinned hard-top, however, had a glass rear window. The soft-top was easy to erect or fold away, but the fitting or removal of the hard-top required two pairs of hands. Until February 1956 the shapely hard-top was

A touring scene from 1959. The 190SL may not have been a pukka sports racer but, with excellent **handling qualities, it was just about in its element on a twisting alpine pass...**

made of aluminium alloy, which made it a good deal lighter than the later hard-tops made of steel.

The interior was deliberately similar to the Gullwing's. Covered in hard-wearing Mercedes-Benz Tex vinyl, or leather as an extra-cost option, the two bucket-type seats were comfortable, most generously padded and fully adjustable. The footwells were covered in plyboard with rubber mats fitted on top originally, but the majority of owners today have quite understandably had this area carpeted as a matter of course during restoration. Behind the seats, the small luggage area could be optionally fitted with a transverse 'jump' seat for occasional use by a third passenger. The top and bottom of the dashboard, the doors and the rear quarter panels and side kick panels were all covered in Tex or leather to match the seat

upholstery. Long armrests with useful integral stowage bins were fitted to the interior door panels

Typically for this period, the dashboard was metal painted in body colour and the instruments, with their black faces and white characters, were a model of clarity. Sitting high up in front of the driver, the main instrument binnacle contained the tachometer and speedometer, the latter calibrated to either 210kph or 140mph. Oil pressure and water temperature gauges sat below the two principal instruments, and after May 1956 a neat circular clock was integrated into the glovebox lid.

The two-spoke steering wheel was typically large, and the horn ring doubled as a switch for the indicators. As a radio – usually a Becker or Blaupunkt – was always an extra-cost option, a removable blanking plate with a chromed 190SL badge sat in the middle of the dashboard. All the switchgear was placed above the bottom bar of the dashboard, but there were so many unmarked switches and knobs that remembering which does what is not especially easy, even for seasoned campaigners.

Developed in complete secrecy, the 190SL prototype was more aggressively styled than the production version.

Besides the obviously different radiator grille, note the bonnet scoop, rectangular sidelamps and rear wing shape.

This Mercedes had a heater and fresh air blower as standard (by no means usual in the 1950s), both being operated by slide controls on the left, centre and right of the facia panel. In period fashion for a 'rag-top', the chromed rear-view mirror sat on top of the facia panel rather than at the top of the windscreen.

Apart from the size of the steering wheel, the only other incongruous aspect of the interior of this sporting pretender was the umbrella 'twist-and-pull' handbrake lever under the dashboard. It performed its role competently enough, but it was not especially conducive to performing a quick handbrake turn! For right-hand drive versions the dashboard was reversed, except that the tachometer remained to the left of the speedometer. Overall, the cabin provides a user-friendly environment: it is a comfortable place to be even by modern standards.

The architecture of the four-cylinder engine, which displaced 1897cc (115.8cu in) and developed 105bhp (DIN), was the same as the Gullwing's six-cylinder engine in that it comprised an alloy cylinder head, cast iron block and alloy crankcase-cum-sump unit. Unlike the 300SL's engine, which was installed at an angle to allow for a low bonnet line, the 190SL unit was positioned in a conventional vertical plane.

In place of the Gullwing's direct fuel injection, the 190SL had twin Solex 44PHH carburettors which were so notoriously difficult to tune that many owners – and even trained Daimler-Benz mechanics – gave up and

fitted Webers instead. For the sake of retaining originality, though, it must be worth persevering with the Solexes these days. Despite the considerable volume of space occupied by the large air cleaner and induction pipe, there is plenty of room for carrying out maintenance.

Amazingly, the crankshaft was seated in just three main bearings, but plenty of really high-mileage examples have proved that this was not a disadvantage even for engines regularly taken up to 6000rpm. The single overhead camshaft was chain-driven from the crankshaft and operated two valves per cylinder via finger type followers. A reprofiled 'sports' camshaft replaced the standard type from as early as July 1955, while another early change was to raise the compression ratio from 8.5:1

Doorless and kitted out in competition guise – Mercedes anticipated the model's use in club racing – with Perspex aeroscreen and special seats, the same prototype reveals its different instrument layout.

Under-bonnet differences between the prototype (above) and production (above right) versions include the 12-volt battery's location and the shape of the induction piping to the carburettors.

to 8.8:1 in September 1956 – the increasing quality of fuels throughout Europe by this time allowed this minor modification. The distributor was also driven indirectly by the crankshaft via a shaft and helical gears, and there was a mechanical fuel pump on the bottom left-hand side of the cylinder block, close to the front.

This power plant was tidily summed up by John Bolster, writing for *Autosport* in November 1955: 'It is as well that this car has such a good gearbox, for the engine demands its constant use. It is not a particularly flexible unit, preferring to turn over briskly all the time. The performance is very creditable, especially when one reminds oneself that the capacity is only 1.9 litres. Nevertheless, something must be sacrificed if you are going to get a quart from a pint pot, and so that delightful gearbox must be called upon to allow the high-efficiency engine to give of its best. Many sports car drivers prefer to handle this type of power unit because although it will run quite well for any driver it will only reveal its full potential for the man who likes to make proper use of his gear lever.'

The engine was also commendably quiet – a necessity for all comfortable cruisers – for a four-cylinder, the exhaust system comprising a three-branch manifold (the central one serving two exhaust ports) converging into a single pipe with a silencer box at the centre of the chassis and another transversely mounted at the rear.

Synchromesh featured on all four forward speeds, and the gearbox was mated directly to the engine, drive being taken to the differential by an open propshaft. The single dry-plate clutch was of 200mm (7.87in) diameter and made by Fichtel & Sachs.

At the front the fully independent suspension was to a conventional design with wishbones, coil springs and Fichtel & Sachs telescopic dampers. The rear was also fully independent, with coil springs and telescopic dampers, but to an unconventional low-pivot swing-axle design. The idea of a low pivot point – as opposed to the Gullwing's high-pivot system – was to achieve a lower roll centre than a normal swing axle, such as those fitted to the Volkswagen Beetle and Porsche 356.

Both halves of the axle cases were attached at the pivot point below the differential housing, and the two axles, or driveshafts, articulated about the pivot point. Lowering the roll centre reduced the tendency for the swing axles to 'jack' themselves up under hard cornering, making for safer handling and better roadholding. Whereas the Gullwing's high-pivot system had come in for a caning from virtually all journalists who were lucky enough to have driven one, the 190SL's road manners were almost universally praised.

Steering was by Daimler-Benz's acclaimed recirculating ball system, the steering box being linked to a Pitman arm (drop arm), unequal-length tie rods and a Stabilus hydraulic steering damper. The majority of road testers considered the steering to be fairly precise but a little heavy, which was especially noticeable when negotiating twisting Alpine passes. In my experience, however, the steering on most cars – even modern ones with power assistance – becomes wearying after tackling a pass like the Stelvio, which has no fewer than 47 tight hairpin bends in quick succession.

Like the later Pagoda-roof cars, the 190SL was always fitted with steel disc wheels; unlike British sports cars, wire wheels were never offered. The wheels, taken from the Sindelfingen parts bin, were the same as the ones fitted to the 180 saloon and shod with 6.40-13 Continental crossplies. Sensibly, the majority of owners these days have switched to modern radials, as these give

With the high-quality soft-top erected, the convertible is almost indistinguishable from the hard-top. Cabin view shows characteristic 1950s-style white steering wheel and luggage shelf behind the deeply-cushioned seats. Superb VDO dials are easy to read at a glance – the speedometer is marked to 140mph on US and UK versions.

better roadholding, safer handling and last longer. Low-profile tyres are not to be recommended, of course, as they put too much stress on the wheel bearings and suspension components.

Unlike the contemporary Porsche 356, the 190SL was fitted with 12-volt electrics right from the start of production, the 56Ah battery being safely sited out of harm's way on the right-hand side of the bulkhead at the rear of the engine compartment.

As early as April 1957, *Sports Car World*'s road testers found the 190SL to be an ideal road car, which handled better than a 300SL Coupé but was in their view 'too

Early cars, like this one, had the number plate lamps mounted on the rear panel, but after July 1957 they were moved to the bumper overriders. Tidy engine bay shows alloy cambox cover and magnificent twin Solex carburettors. Famous three-pointed star and 190SL badge, both in high-quality chrome, sit proudly above the lock on the alloy boot lid.

slow for competition work'. However, the report's author also comments: 'Daimler-Benz, unlike other manufacturers, is willing to admit that a car is occasionally a cranky device, and that it will perform better and last longer when you face the fact and take appropriate steps.' The 190SL's owner's manual carries stern advice against ignoring the rigorous servicing schedule and, as *Sports Car World* noted, 'The theory behind all these directions and the elaborately worked-out maintenance programme is that, if the owner will follow them, he'll never have a moment's trouble with his car. It's more than just pleasant; it's habit forming.'

This just about sums up what the 190SL is all about – maintain one as Daimler-Benz originally intended and it really will go on for ever. Treat it like a racing car, on the other hand, and sooner or later it will act like one – usually with a bang followed by expensive repair bills.

Production modifications.................

As a result of Daimler-Benz's deliberate policy of carefully refining the 190SL down the years, rather than making sweeping changes, it may well appear to the untrained eye that it remained in its original form throughout. This is,

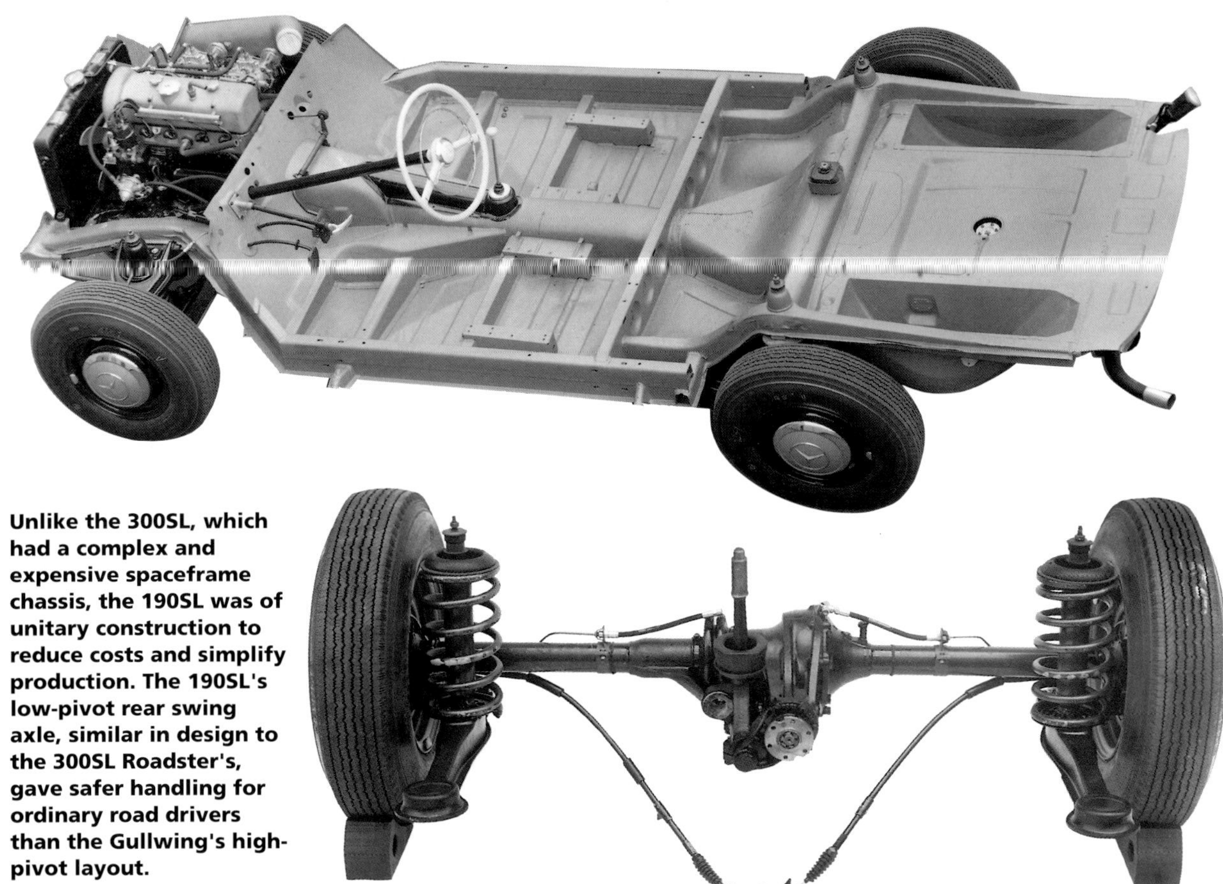

Unlike the 300SL, which had a complex and expensive spaceframe chassis, the 190SL was of unitary construction to reduce costs and simplify production. The 190SL's low-pivot rear swing axle, similar in design to the 300SL Roadster's, gave safer handling for ordinary road drivers than the Gullwing's high-pivot layout.

of course, far from the case, as dozens of modifications were made every year.

Many of these were relatively insignificant and even went unnoticed by owners. For example, an early change involved the cylinder head gasket, which was increased in thickness in April 1955 from 1.5mm to 1.7mm. The difference was almost visually imperceptible but, without doubt, the engineers in Sindelfingen had good reasons for making this change.

A more noticeable change occurred in April 1955 when the axle ratio was increased from 3.70:1 to 3.89:1. Reinforcing ribs were built into the crankcase from August of the same year. Interestingly, the rear axle ratio was changed again, albeit less significantly, in September 1955 from 3.89:1 to 3.90:1. Just three months later, the wheelarch 'eyebrows' were fitted with bright mouldings as standard on cars ordered with a hard-top, but this decoration was an extra-cost option on roadster versions. Like most pieces of body trim, these have a tendency to trap moisture and cause corrosion.

At the start of 1956 the roadster's seat backrests were made 20mm (0.79in) narrower and were less comfortable as a result, but if journalists noticed this change they certainly did not comment upon it in their road tests. At the same time the accelerator pedal, which had previously been left bare, was fitted with a rubber cover for

enhanced grip. Considering Daimler-Benz's meticulous attention to detail, it seems odd that a foot pad had not been fitted from the start of production.

A second Bosch horn and an ATE brake servo were standardised from April 1956 and, in the middle of the same year, the larger and more easily visible tail lamps from the 220 saloon found their way onto the hind quarters. From the beginning the 190SL had an exterior mirror on the driver's side, an additional one for the passenger's side being an extra-cost option, but cars produced for the home market were treated to a mirror on each side as standard from November 1956.

Just a month later the alloy bonnet and boot lid were increased in thickness from 1.25mm to 1.5mm, in order to add strength to what was hitherto considered to be relatively flimsy panelwork – something that became readily apparent with the lids propped open in windy weather. Bent aluminium made by Daimler-Benz was not especially desirable when one considers the substantial cost of replacements…

The production engineers had several field days with 'widgetry' during 1957. In January, for example, the cork fuel tank float was changed for a plastic one, and, more visibly, the tail lamps were made convex at the same time. The interior door handles were modified in February. In July the clutch lining on the flywheel side was changed

from Rusko to Klinger material, and the rear number plate lights were moved from housings on the outside edges of the number plate itself to the bumper overriders. Towards the end of the year 190SLs exported to North America began to differ a little from European-specification cars, having red indicator lenses instead of amber from October and indicators of the self-cancelling variety from November.

At the start of 1958 US cars also lost the dividing strip between the stop and tail lights. Like early Volkswagens and Porsches, the 190SL initially had rather elegant sun visors in tinted Perspex with a high quality chromed framework, but for safety reasons these were changed in May 1958 for padded visors covered with leather or Tex. Interestingly, this

coincided with a similar change for both the Beetle and Porsche 356.

It was a small irritation that the majority of German cars during the 1950s, and beyond in many cases, came with separate keys for the ignition, boot lid and fuel filler cap, but this was addressed early in 1959 when one key sufficed for all functions. In April the optional fog lamps were fitted with extended eyebrows, and headlamps with asymmetrical beams were introduced at the same time for export markets.

Another minor point, but one which may help in dating a particular car, is that the tiny dust cover fitted over the top of the glovebox lid lock did not appear after July 1959; windscreen washers were fitted as standard at roughly the same time. From the end of September the passenger's sun visor was

190 SL
444 BJF

The 190SL's aerodynamically sound body shape (opposite) was in keeping with contemporary thinking, and aided the car's relatively high top speed of around 105mph. The 190SL's glamour was irresistible to one movie maker: this is Grace Kelly and Frank Sinatra in *High Society*.

omitted on the hard-top, and the Roadster's visor was fitted with a small vanity mirror.

By the end of the decade traffic had increased markedly and Daimler-Benz acknowledged this by increasing the size of the rear windows on both the hard-top and soft-top from October 1959. The window on the hard-top was particularly impressive, as the new glass wrapped around the rear quarters of the roof to give faultless rearward vision. Apart from improving safety, this change let more light into the cabin and gave a bright airy feel to the interior, although arguably it detracted a little from its essentially snug character.

From November 1959 a right-hand reversing lamp was fitted as standard, a feature of American-specification cars from the beginning of production. This was yet another small but significant modification.

In its 1960 test *Road & Track*'s report comments: 'As a tribute to the excellent all-round, well-balanced design, no really large changes have been made since our first test of the car... Beginning in 1960, the optional hard-top was given a new shape more in conformance with that of its larger running mate, the 300SL. Other than that there have been no changes, nor will there be any changes for 1961.' The writer of this piece was wrong in his assumption, for improvements continued to be made until the end of production.

From February 1960 the headlamp flashers became more powerful as the 15w bulbs were changed for 18w

units. Bomoro double-wedge door locks were introduced in April. In August the boot lid handle was enlarged so that it was easier to grip, but the new design was considerably less elegant than the chromed finger latch it replaced. From November a modified Bosch VJR BR 24T distributor was fitted. Late in December the Bosch headlamp housings became galvanised, a change that attempted to delay the almost inevitable rusting process.

Throughout 1961 the car continued to be modified and, as before, the vast majority of changes were relatively trivial. At the beginning of the year, the fuel filler cap received a Mercedes star motif and an inscription, 'Ohne Luftung' (without ventilation). At the same time the interior heating and ventilation operating handles were made from softer Hostalen plastic, a safety measure aimed at reducing the possibility of serious injury in the event of a collision. Seat belts would have been a more worthwhile feature, but anchorage points were not fitted until October 1961.

A few minor engine changes were made during 1961 and included stronger valve inserts from May, screw plugs replacing the previous cylinder block expansion plugs from June, and a stronger crankcase gallery from July. A groove was cut into the crankshaft's front bearing from August in an attempt to eliminate a sucking noise, which had caused complaints from customers from the beginning, and at the end of the same month the

This 1960 190SL reveals itself as a late car by its hard-top style – with wrap-around rear window – and overrider-mounted number plate lamps. This car lacks front bumper overriders, which were an extra-cost option.

crankshaft received hardened circumferences on the first and fourth main bearings and connecting rod journals.

For 1962, when Daimler–Benz's sales figures for the 190SL – 2246 in total – were predictably starting to tail off, the 'baby' sports car was treated to just 14 under-the-skin modifications, but the engineers at Sindelfingen obviously considered them to be worthwhile. From the middle of January, the gearbox was slightly modified in that the tooth width on second gear was increased from 16mm to 17mm, and on third gear reduced from 20.5mm to 19.5mm. At the end of the month the mounting holes for the seat belts on the transmission tunnel were usefully closed with sealing plugs, which naturally were dispensed with if seat belts were fitted.

Textar V643 brake linings were fitted in March and the handbrake lever was reinforced. A new Bosch VJR 4

Intended for very occasional use, this rear 'jump seat' (right) was an early extra-cost option. In the absence of a 'jump seat' (far right), luggage space behind the tipping seat backrests is generous, as befits a sports tourer. Note the hard-top's luxurious headlining. Alloy plate on left-hand lower A-pillar gives chassis number and paint code.

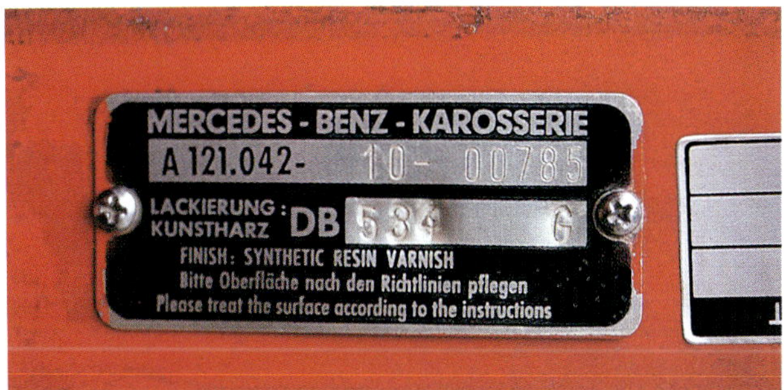

BR 29T distributor was introduced in April, and alternative pistons by Nural and Mahle were fitted from May. Until this time motor manufacturers had taken very little trouble to treat their cars against corrosion, but Daimler-Benz took the important and far-sighted step of spraying PVC under the front wheel arches – but not the rears – from May.

The optional fog lamps were fitted with yellow reflectors instead of yellow glass from July, and from September the brake linings were changed yet again, from Textar to John Manville items. From October the chain tensioner housing on the front of the engine was modified and, at the end of November, a gasket was fitted between the gearbox housing and cover.

Between 1955-62 motoring writers penned thousands of column inches about the 190SL, and each road report throughout production differed little in tone and context overall. Each praised the build quality and paint finish, handling and performance, although the majority were agreed that the car was not race track material. Recurring criticism was aimed at the heavy steering (a problem that was never addressed), the over-large steering wheel, an optimistic speedometer, and the constant need to change gear to get the best from the 1.9-litre engine. Enthusiastic drivers, of course, did not regard the latter as a disadvantage: shifting gear and over-use of the right foot

on the throttle pedal is, after all, a lot of fun and part of the inherent nature of any sports car, whether it is a 'real' or 'pretend' one.

As *Road & Track* concluded in its December 1960 test: 'This list price of $5129 quoted is a bare car; local taxes and license are extra, of course. Our test car had the expensive Becker Mexico radio, genuine leather upholstery and both tops. It lists for $5758 so equipped. This seems like, and certainly is, a lot of money. We could dwell at great length over the really superb quality that goes into the products of Mercedes-Benz and it would be true, but boring. The best way to determine value is to look the car over carefully, both inside and out, give it a good test drive and form your own opinion. We say it's well worth the money.'

By the beginning of the 1960s, there was no shortage of sports cars on the world market and many, particularly from Britain, offered better performance than the 190SL, and at prices a good many more people could afford to pay. The 190SL was seen by some, therefore, as something of an enigma.

In its December 1961 road test *Sports Car Graphic* summed up the 190SL rather well: 'Although we've been enthusiastic over several Mercedes models and lavished well-deserved praise on them, the only way we find the 190SL even likeable is if we constantly rationalise that it is

The earlier hard-top style seen in a contemporary press shot – a larger rear window was introduced in October 1959.

a prestige 'personal car', no more, no less. A description for it that one knowledgeable person advanced was: "An old man's roadster". This perhaps is a bit severe but does accurately nail down the car's concept. It's been three years since we road tested the 190SL and, though there have been many improvements, our impressions remain unaltered; it's a dull-but-expensive piece of transportation in sports car clothing. While this opinion may infuriate some proud 190 owners, we doubt it will even raise an eyebrow at the factory, as we suspect this is exactly what they had in mind.'

This last comment demonstrates how the 190SL came to be regarded, which is broadly why, despite the constant fine tuning, the 190SL was replaced in 1963 only for the same process to start all over again – as it always has done, of course.

Driving impressions ....

Most drivers who have exploited the 190SL to its full extent over many thousands of miles – and this is the only true way to get to know a car well – will rarely admit to having felt like the king of the highway. They may privately confess, though, to having discovered what it is like to be the prince of suburban avenues.

It is almost futile to make comparisons between the road performance of the 190SL and other sporting cars because, in many respects, the Mercedes was without a direct rival. Alfa Romeo's 2000 2+2 Spider, launched in 1958, was comparable in that it had a four-cylinder engine developing 115bhp, but this classic twin-cam gave better performance – up to 112mph (180kph) – and was more in the nature of a true sporting car, which meant it was lighter, less well appointed and less expensive.

The pushrod-engined 356 Porsches performed as well from rest to the benchmark figure of 60mph and had a similar top speed, but the difference between the two

Stuttgart cars was that Porsches achieved these figures with significantly smaller engines. Contemporary Ferraris, always with V12 engines, do not even come into the equation, while the Aston Martin DB2/4, launched in 1955, was an altogether different car in the mould of a truly sporting British thoroughbred.

Closest in concept, comfort and performance were the bespoke 2-litre straight-six Bristol 404 and 405 saloons and drophead coupés, which were usually good for around 110mph (177kph), but these hand-built carriages were aimed at a wholly different clientele and made in such relatively small numbers that direct comparison with the 190SL serves very little purpose. So, the 190SL forged a path of its own.

Jump behind the wheel and it is quickly obvious that the wide, spacious cabin is comfortable, the seats beautifully upholstered. Although the seats lack lateral support, they offer plenty in the way of back and thigh support to minimise the possibility of muscle fatigue on long journeys. Vision through the panoramic windscreen is good, and the commendably narrow pillars reduce the possibility of an obscured view when, for example, pulling out of a T-junction.

Luggage space is generous enough for a two-seater, with sufficient room behind the seats for a bag or two and various odds and ends, and a boot that, although shallow, is sufficiently roomy for a couple of suitcases. In many respects, therefore, the 190SL makes an almost ideal touring car.

As one might expect, the Bosch electrical system has proved to be totally reliable over the years, which is why a flick of the ignition key is usually sufficient to engage the starter motor instantly and fire the power unit into life. Pulling away from rest is undramatic, as the clutch if properly adjusted, feels nicely weighted and silky smooth, and the gearbox works with precision. No-one has made a gearbox that comes close to the Porsche unit fitted to the 356, but this Daimler-Benz 'box has been judged by many to be the next best.

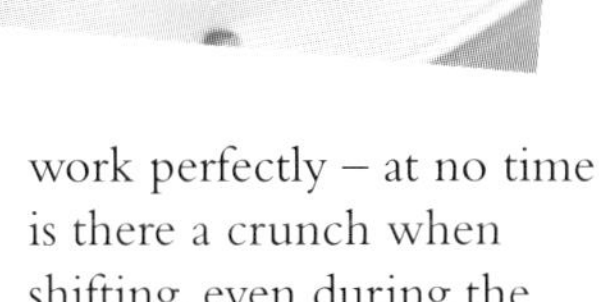

work perfectly – at no time is there a crunch when shifting, even during the most energetic use.'

The interesting observation that top gear is on the tall side was undoubtedly caused by the writer's understandable inability to appreciate that all German cars of this era were the same – they had to be to cope with constantly high engine speeds on long, straight *autobahns*. By today's standards the power unit is not especially smooth at idle, but some journalists reckoned it to be one of the quietest engines of the time, to the extent that on the move it is not always possible to tell that it has only four cylinders.

For absolute maximum performance, it is always necessary to use 6000rpm in every gear and change up as quickly as possible. Few owners who knew what was good for their bank balance did this on a regular basis, but journalists felt obliged to for the sake of making comparisons with other cars, and to answer the perennial question of 'what'll she do, mister?' from teenage boys. Typically, 0–30mph came up in 3.8sec, 0–60mph in 11.5sec, 30–50mph (in second gear) took around 4.7sec and 50–80mph (using second and third) could be achieved in 13.2sec. By today's standards these figures are reasonable respectable, although hardly sizzling, but the car acquits itself well when measured against some rivals.

In its October 1955 test *Road & Track* noted: 'The only criticism which might be made by the non-enthusiast is that the car feels high-geared in 4th, and 3rd is necessary for rapid ascents of long steep grades. About 20mph is the minimum in 4th gear, and we used 3rd gear at all times below 35mph. The transmission is, incidentally, one of the best. It is nearly dead silent in all ratios and the synchronisers (on all four forward speeds)

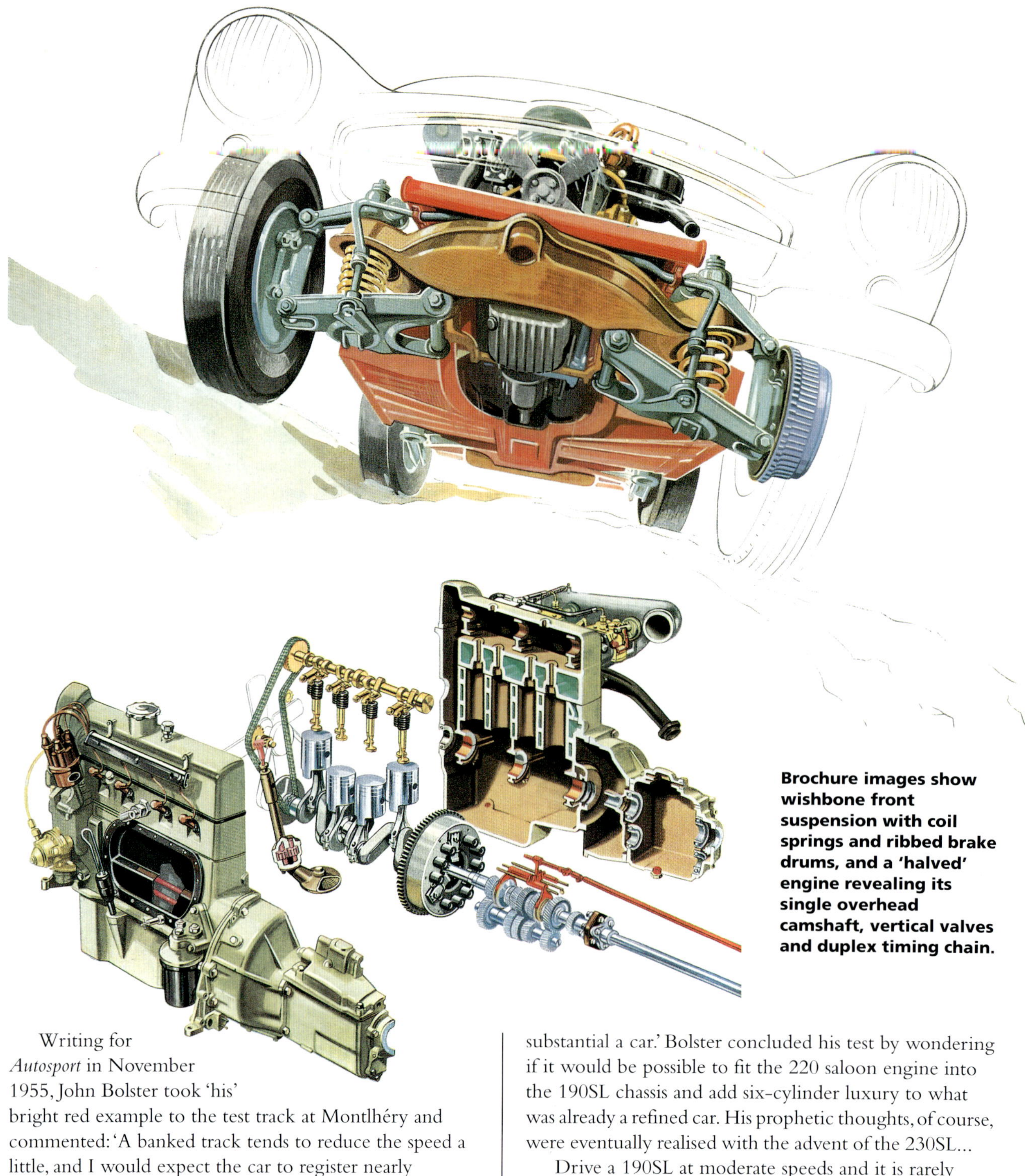

Brochure images show wishbone front suspension with coil springs and ribbed brake drums, and a 'halved' engine revealing its single overhead camshaft, vertical valves and duplex timing chain.

Writing for *Autosport* in November 1955, John Bolster took 'his' bright red example to the test track at Montlhéry and commented: 'A banked track tends to reduce the speed a little, and I would expect the car to register nearly 100mph on my usual dead level piece of road. The speedometer showed only a very moderate optimism, and I had it round to the 180kph mark on occasion. The acceleration figures reveal a lively performance, and the 0–60mph time of 11.2sec is particularly noteworthy for so

substantial a car.' Bolster concluded his test by wondering if it would be possible to fit the 220 saloon engine into the 190SL chassis and add six-cylinder luxury to what was already a refined car. His prophetic thoughts, of course, were eventually realised with the advent of the 230SL...

Drive a 190SL at moderate speeds and it is rarely anything more or less than a dull motoring experience, but pick it up by the scruff of the neck, boot it hard, and its finer points soon become readily apparent. As Walt Woron discovered in his test for *Motor Trend* in December 1955, 'When the road begins to wind you get the general

Few contemporary journalists could fault the 190SL's road manners, and such is the car's build quality and engineering integrity that it still feels great to drive today.

feeling that the 190SL will handle like a sports car, but don't let this lead you to a wrong assumption. As you speed up and begin to push it into the corners, you get a sort of sloshing around (the tires seem to roll under and the car appears to shift sideways). On a real sharp (90 degree) turn taken fast (30mph) for its radius, the back end will break, but the steering is so positive you can easily correct without even changing the throttle opening…On faster (40mph and up) turns, you can take it through in an easily controlled four-wheel drift. Body lean is noticeable but not objectionable to the driver, requiring the passenger to hold on.'

For those who switched from Porsche ownership – and significant numbers did – there was one particularly distinguishing feature between the two cars. The Mercedes continued to travel virtually arrow-straight in gusting side winds, whereas the 356's rear weight bias required constant corrections with the steering wheel.

The really great feature of the rear swing axle suspension was that it always gave notice that the limit of tyre adhesion was approaching by flicking from understeer to oversteer. Those who were afraid of it and never learned the technique needed to drive a 190SL (or any other car with swing axles) condemned it almost to a man. In reality the critics were actually demonstrating publicly more about their driving ability – and more particularly, their lack of it – when they aired their feelings about 'vicious' oversteer.

The truth was understood by *Sports Car World*, which noted in its April 1957 test: 'Where the 190SL chassis really shines is in roadholding on curves and straightaways at high speeds. It squashes close to the ground and tracks true on the straights like the 300SL. It stays glued in the turns better and its resistance to power slides is far, far greater. You'd have to be driving well beyond the limits of common sense to get into trouble with a 190SL.'

Naturally, the vast majority of journalists and owners alike never reached the limit of adhesion, let alone

Daimler-Benz's publicity shots often emphasised the 190SL's touring nature, evoking a world of leisure, fun and adventure. Here we see a picturesque harbour-side scene, a happy meeting on the open road and a holidaying 190SL owner motoring through the streets of Pisa.

exceeded it, but by the end of the 1950s journalists had fallen into the habit, regrettably, of repeating in parrot-fashion most of the drivel they had read in previous tests. This kind of reporting is reprehensible in the extreme, but fortunately has largely been eradicated from the better quality magazines today. And to be fair to the Mercedes, it was shod with crossply tyres, which never helped to create a good impression of the car's capabilities through the twisty bits.

The 190SL's ability to stop was generally on a par with its performance. In normal driving conditions the large drums work well enough, but many discovered the horrors of 'fade' after repeated, hard use on mountain roads, some reporting that the pedal had a disconcerting habit of travelling to the floor on occasions. As the drums were made of cast iron, it is best to assume that the claim in *The Autocar*'s 1958 road test that they were of 'light-alloy' was the result of the writer having an 'off' day.

During the time that has elapsed since the 190SL went out of production, improvements in braking and tyre technology have been dramatic, which is why great care must be exercised when driving cars like the SL on today's roads. I have driven the only 300SL Roadster in the world fitted with an ABS braking system: whereas it felt modern and lively in most respects, the car's stopping

ability remained, nevertheless, the one facet that dated it. Although the 190SL is incapable of reaching the 300SL's heady heights of more than 130mph (210kph), having to haul down the 'baby' Merc from 100mph in a hurry would offer food for thought, especially on a wet road.

With the hard-top in place the car feels like a regular coupé, of course, but the soft-top is so well made that it is sometimes easy to forget you are driving a convertible. The soft-top can be erected or folded away by one person very quickly, as it is clamped to the windscreen pillars on each side by beautifully chromed latches. Weather protection is good even in quite heavy downpours but, as you might expect, there is always going to be some wind buffeting with the hood up and the frameless side windows lowered.

The Autocar concluded its 1958 test with the following comments: 'The 190SL was approached with keen expectation, knowing the reputation of its makers for quality and workmanship. It proved to be fast and tireless, exhilarating to drive, and was obviously created with long-distance, comfortable travel in mind rather than competition work. Naturally its powers of acceleration do not approach those of the larger engined SL type. In all aspects of handling and control it is certainly one of the safest cars tested by this journal.'

Over a total distance of 1187 miles covered by *The Autocar*'s staff, the 190SL recorded fuel consumption of 22.4mpg – impressive considering that much of their mileage was in the mountains of continental Europe. The

list price in Britain was quoted at £2896 at this time, or nearly five times as much as a Volkswagen Beetle 1200.

John Bolster, incidentally, was not the only one who noticed the optimism of the 190SL's speedometer. In 1958 *Sports Cars Illustrated*'s John Christy commented that this was the only, 'real complaint we could find on the car', and added, 'With everything else so letter-perfect this relatively minor fault becomes almost glaring. True, it does allow one to apparently nudge the legal speed limit and still avoid summonses but, for all other purposes, it requires a sort of mental conversion as though it had been calibrated in kilometres instead of miles.'

Criticising the speedometer for inaccuracy really demonstrates how most writers really struggled to discredit the 190SL in any meaningful way. On reflection, there are plenty of features which, purely from a driving point of view, date the car. In most modern cars, for example, there is a console – mostly containing switchgear and fresh air/heating vents – rising from the centre of the floor to the dashboard, and separating the driver and passenger footwells. In period fashion, the 190SL has nothing other than a shallow transmission tunnel to create a division between the footwells.

In conclusion the 190SL was and remains a competent tourer. Although not a driver's car in the accepted sporting sense, it was entirely what Daimler-Benz intended it to be. While some criticised it for its lacklustre performance, the engineers and designers in Stuttgart kept confidently quiet, resolutely refusing to respond to the whims of motoring journalists.

Sports Car Graphic's 1961 road test hit the nail on the head when its author summed up the car in this way: 'The car's biggest success is its public impression, even though it's been around long enough to be fairly common in appearance. Driving a 190SL, you LOOK like a sport and we had to fight off one member of another Petersen publication who constantly wanted to borrow the car for girl-baiting tours down Sunset Boulevard. Which, if we may say so, is indulging in what can already be an expensive hobby.'

THE PAGODAS

Instantly nicknamed 'Pagoda' after the roof profile of the hard-top, the 230SL, with its crisp styling, was a dramatic step forward from the first-generation SL. The visibility afforded by such large windows was a great aid to safety.

The launch of the second-generation SL in 1963 ushered in a new era of sports car design, with Daimler-Benz at the forefront of the gradual redefinition of the concept of sports cars that occurred during the decade. The traditional image of a sports car – separate and comparatively ill-handling chassis, noisy 'fire-breathing' engine, flimsy single-skinned soft-top that leaked – gave way to many designs that were more comfortable, considerably safer, handled exceptionally well and were still reasonably fast.

What some of the new-generation sports cars lacked in straight-line speed – an increasingly irrelevant factor by this time – was more than made up for with a chassis that really could negotiate corners without frightening drivers of average ability. As each manufacturer strove to improve engineering standards, sporting cars changed out of all recognition, some of them, such as the Aston Martin DB6, becoming sophisticates with hitherto unknown levels of equipment, catering for customers with increasingly large bank balances.

The 230SL was the epitome of this trend towards a new kind of sports car. The SL badge did not really stand for *Super Leicht* (Super Light) any longer, as 'Super Luxury' was much more accurate and to the point.

As with past models in Daimler-Benz's armoury, the 230SL was largely a product of Rudi Uhlenhaut's brilliance and the car reflected this perfectly. As the writer of *Autocar*'s 1963 230SL road test made clear, 'It is my firm belief that an outstanding engineer can impose his personality on a creation even in these days of large establishments. Uhlenhaut seems to be able to do this yet control a department of 380 people, of which 120 are designers and 160 qualified development engineers. By organising his day-to-day routine so that he maintains close personal contact with all development work and doing his reading of technical reports at home, he

Extensive pre-launch testing confirmed the 230SL's peerless handling, and where better to evaluate this than at the 14-mile Nürburgring? Styling was the work of Paul Bracq, who used his abilities with the brush to paint this retrospective portrait of his masterpiece many years later.

in 1963 after a driving exercise with him, 'Many drivers during these two periods soon learned that it was impossible to mislead him about the car's behaviour in any way; he could jump into the car, lap all circuits as fast as most of them, and analyse a particular facet better than any. Today, at 55 years of age, he seems to have lost little of his driving ability and none of his enthusiasm.'

Following on from the design of the 190SL, the new 230SL 'Pagoda' had a conventional monocoque hull which, in contrast to the 300SL's multi tube spaceframe chassis, was considerably less complex and relatively cheap to build. Steel was used for the main structure of the bodyshell but, like all previous SLs, the doors, bonnet and boot lid were made from aluminium alloy to save weight. For ease of manufacture the chassis pan was of relatively simple construction and immensely strong, basically comprising a central tunnel to clear the gearbox and propshaft, two floor sections, box-section sill panels and sturdy longitudinal members on each side of the transmission tunnel. These longitudinal members formed a forked frame which ran to the front of the car and was joined by a large crossmember that carried the front suspension. At the rear there were separate box sections that curved upwards to clear the rear axle assembly and continued along the sides of the boot floor.

Most of those who were present at the car's Geneva Show launch did not give a hang about such things; the

manages to cover many thousands of miles per year in experimental vehicles and always makes the last pronouncement on that largely subjective characteristic – handling.'

Uhlenhaut joined Daimler-Benz in 1931 as a technical assistant; needless to say, his outstanding ability was quickly recognised and utilised to the full. When the factory went racing with its 'Silver Arrows' Grand Prix machines in the 1930s, Uhlenhaut was put in charge of the competitions department. He was responsible for developing the cars and they were fabulously successful, remaining legends of motor racing to this day.

When Daimler-Benz resumed motor racing in 1952, Uhlenhaut's talents came to the fore yet again and the quite amazing success continued. As *Autocar* pointed out

Early attempt at a new mid-sized SL – a car to replace both the 190SL and 300SL – produced this 'halfway house' proposal (left), which has Pagoda hints but retains some curvature in its shape – the date displayed by the front wheel is 13 December 1959. Much nearer the real thing is this pre-production prototype (below) photographed in March 1963 – before production began the Mercedes-Benz script on the front wing was deleted and wheel diameter increased from 13in to 14in.

vast majority were more interested in the car's looks and its interior fittings. Above all else, it was the body styling that initially created the real interest, even if the car's looks could not quite match the sensation that the Gullwing had caused nine years earlier. A squat, wide stance gave added purpose to the crisp, lean shape, making the 230SL such a complete contrast to the big, heavy saloons that formed the rest of the Mercedes range.

In typical Daimler-Benz fashion, the 230SL was just as solidly constructed and 'over-engineered' as any of the saloons, but at least it looked the part of a lithe and athletically fit sports machine. However, not everyone held this view. Writing in *World's Fastest Sports Cars* in 1965, David Owen was not especially impressed with the styling and commented: 'The car isn't particularly remarkable to look at. Unlike many all-look-and-no-go sports cars, it looks like a rather scaled-down sedan, especially with the rather Chinese pagoda-shaped hard-top fitted. The styling is reasonably neat and tidy, without the sharp brilliance of the better Italian cars, but the vast wide track, the fat tyres and low crouch make the car as squat and purposeful as a bulldog or a Sumo wrestler; this isn't just to impress. This is the secret of the car's success.'

The interior was as beautifully manufactured and screwed together as every other part of the car. In contrast to so many sports cars of the period, it was also ergonomically excellent – just one reason why it changed so very little down the years. Unlike the 190SL, the generously padded seats gave good lateral support and, as in the past, Mercedes-Benz's high-quality Tex vinyl – or leather at extra cost – was used for the upholstery. Although space behind the seats was limited, there was sufficient room for a reasonable quantity of luggage and, as before, a transverse 'jump' seat for occasional use by a third passenger was offered as an extra-cost option. The floor, sills, transmission tunnel, seat backs and rear luggage space were all generously carpeted in a typically hard-wearing woolcloth material.

One hugely improved area was the dashboard, which, unlike that of the 190SL, was mercifully uncluttered with excessive switchgear. As usual, the dials were simple, crystal clear and easy to read at a glance through the large steering wheel. In fact the layout of the facia panel and instrumentation was not too dissimilar to the 300SL Roadster's: there was a large hooded binnacle with the speedometer on the left, the tachometer on the right and

Two shots from a styling 'sign-off' presentation early in 1963. A pre-production prototype sits on a studio turntable in company with two styling models (right), while a member of the styling team demonstrates the wheelarch flares that replaced the distinctive 'eyebrows' of the 190SL (below). It is curious that crossply tyres are fitted: the suspension of the 230SL was specifically designed for radial tyres and the car was never sold with crossplies.

a roughly rectangular combination gauge in the centre containing the fuel and oil pressure gauges and the various warning lights. All the instruments were made by VDO, and further included a circular clock in the centre of the dashboard. Two swivelling fresh air outlets were sited at the extreme ends of the dashboard, with their control slides in the centre above the radio position.

The top and bottom bars of the dashboard were padded and trimmed to match the seat upholstery and interior panels. Wood veneer trim, which has often looked slightly cheapening in post-war Mercedes-Benz cars, was confined to the centre console between the seats, a wide strip following the curvature of the windscreen across the top of the dashboard, and the radio speaker grille on the top of the dashboard. It is most curious that wood veneer – a hangover from vintage days – should have been used because this clearly did not fit in with Uhlenhaut's philosophy of producing a technically advanced car. Wood had been used to good effect, and with greater aesthetic success, by some traditional British manufacturers, but no doubt the SL's incongruous trimmings were appreciated in some quarters.

As always, Uhlenhaut considered that 'his' sports cars should reflect contemporary trends in motor racing – the 1.5-litre Grand Prix formula that was in place between 1961-65 in this case – by using a chassis with the capability to match the engine fitted in it. It was purely on this philosophical basis that the 230SL was conceived and designed: all the individual engineering elements should knit together in a balanced whole.

The role played by the newly invented *Halbgurtel* (half-belted) Continental RA60 radial tyres cannot be over-emphasised. These tyres had specially constructed sidewalls to ensure that they remained 'stiff' and as near to the vertical plane as possible under hard cornering. Early

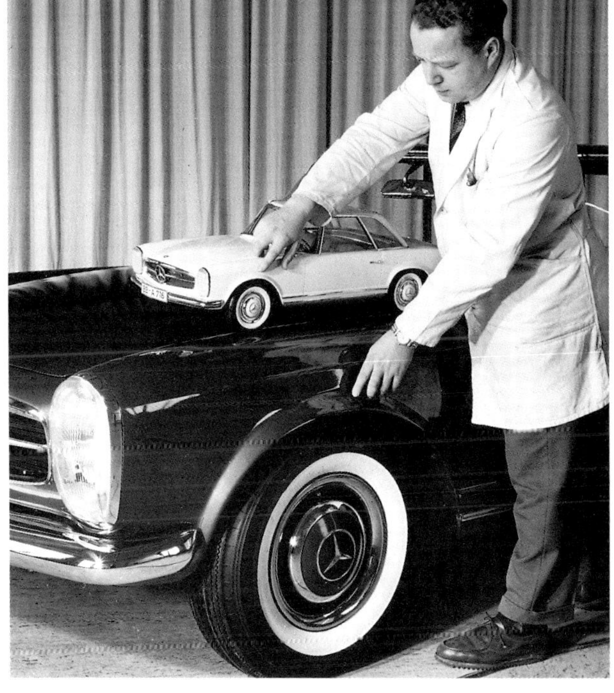

Michelin X radials had proved to be far too flexible when fitted to high-performance sports cars like the 300SL, and tended to distort circumferentially during hard acceleration and deceleration, in much the same way as the enormous tyres fitted to the back of drag racing machines. Because the cords used in the construction of the Continental RA60s were wound at an angle below the outer tread, they were not true radials in the modern sense but they certainly made for an acceptable compromise in considerations such as ride comfort and roadholding.

Uhlenhaut designed the 230SL's suspension to suit the roadholding characteristics of the new tyres and the extra cornering loads that they were known to impose. The front suspension was fairly conventional in utilising double wishbones, coil springs, Bilstein telescopic shock

With soft-top in place, the 230SL not only remained aesthetically sound but also proved to be more aerodynamically efficient than with the hard-top. Head-on view best shows the hard-top's distinctive shape as well as emphasising the car's purposeful stance, with wide track and low ride height. Discreet boot badging is one of very few ways to identify the three Pagoda models externally.

absorbers and a 22mm diameter anti-roll bar (reduced to 20mm on the later 250SL). At the rear Uhlenhaut chose to use a low-pivot swing-axle system similar to the ones employed on both the 190SL and the 300SL Roadster but, by increasing the track substantially, the point at which the axles would 'jack' themselves up would occur much later and at much higher speeds. As at the front, coil springs and Bilstein telescopic dampers were fitted.

A feature first seen on the 300SL Roadster was also adopted for the 230SL's rear suspension. An equaliser – or compensator – spring was fitted horizontally above and between the two halves of the swing axles, to the right of the differential casing, in order further to keep the swing

axles in check, but without affecting the spring rates under roll through the corners. This spring was simple and effective, but was changed for a much more complex hydro-pneumatic device on late 280SLs that, in theory at least, performed much the same job but in a more precise manner. Incidentally, because of the expense and complexity of the hydro-pneumatic 'springs', many late 280SLs have reverted to the earlier mechanical coil spring set-up and appear to be none the worse for it.

Like the 190SL, the 230SL had to sell well and be profitable and, to that end, several components in its make-up were taken directly from other Mercedes models. Although not exactly the same, the in-line six-

By comparison with the first-generation SLs, the 230SL's facia is laudably uncluttered and entirely functional. Switchgear was simplified and instrumentation reduced, but the steering wheel – available in white or black – was still too large. The 230SL's four-bearing engine was noted for its smoothness and ability to rev freely; in this smart engine bay the alloy castings have their original natural finish.

cylinder engine – with cast iron block, alloy cylinder head and steel sump pan – was based on the lump fitted to the 220 range, but modified to give extra power. To begin with the power unit was increased in size from 2195cc (133.9cu in) to 2306cc (140.7in) and the compression ratio raised from 8.7:1 to 9.3:1. The exhaust system was modified and larger valves were installed, but using the saloon's four-bearing crankshaft meant that the free-revving nature of the engine was retained.

The Bosch fuel injection differed in that the 230SL used a direct system, with the injection nozzles mounted in the induction ports, fed by a six-plunger injection pump driven at half engine speed. The 220 saloon, by comparison, had a much simpler two-plunger pump working at full engine speed, with the fuel fed into a nozzle in the aluminium alloy manifold.

The result of all this development was that the 230SL had an altogether smoother and more powerful engine that idled in virtual silence at around 600rpm. However, there was a price to pay, for the engine's enormous complexity put DIY maintenance well out of reach, even in the relatively electronics-free days of 1963, and made professional servicing fairly expensive.

Supplied with fuel from the tank by electric pump, the fuel injection pump, which sat to the left-hand side of the cylinder block, was almost a miniature replica of the

Plan diagram of chassis elements is particularly informative for showing the rear suspension layout. A single trailing arm on each side locates the casings of the swing axles, whose deflection under hard cornering is restrained by a compensator spring in the centre.

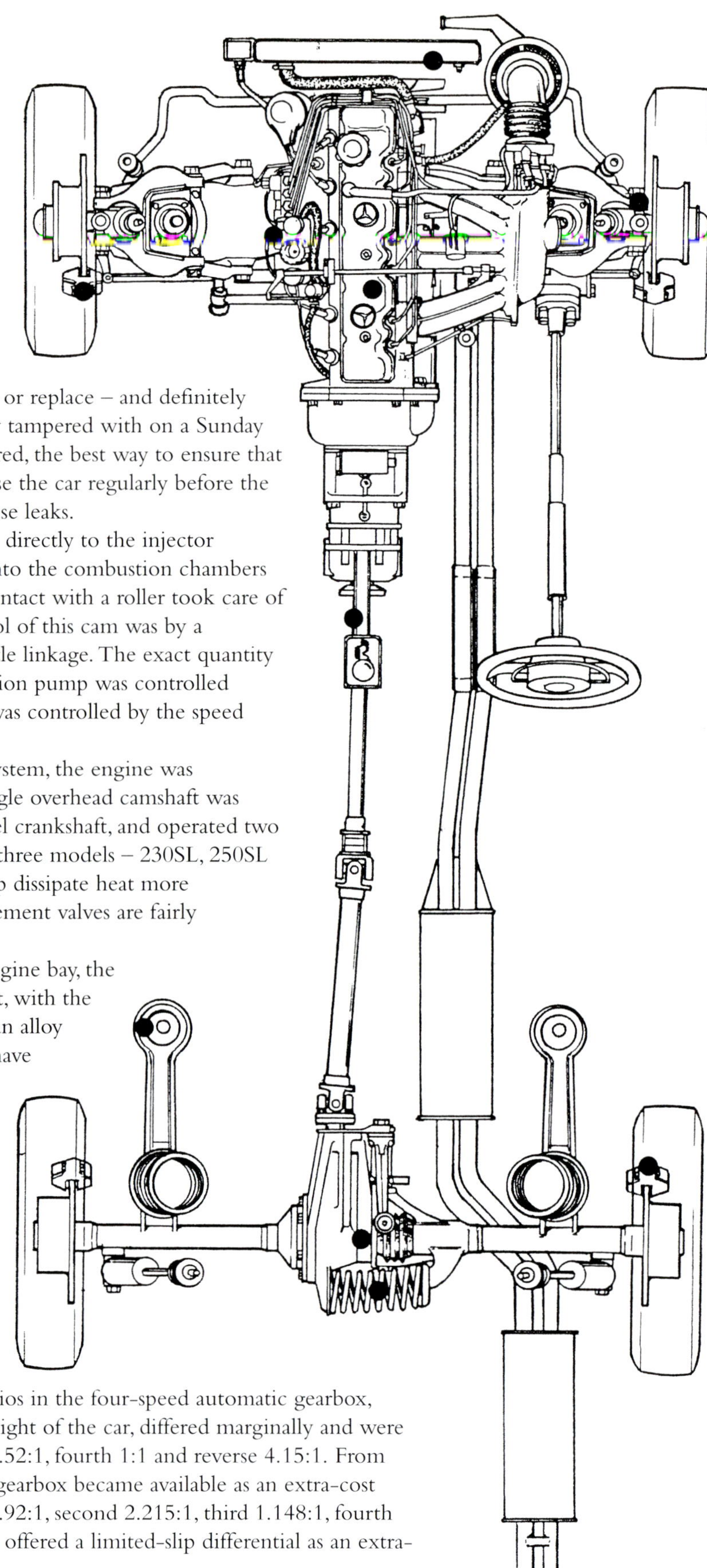

six-cylinder engine which it fed, and as such a complicated and costly item to repair or replace – and definitely not the kind of gadget that could be casually tampered with on a Sunday morning. But, as many owners have discovered, the best way to ensure that the pump never needs to be replaced is to use the car regularly before the various seals get a chance to dry out and cause leaks.

Fuel was supplied by the injection pump directly to the injector nozzles through the fuel lines, and sprayed into the combustion chambers at 213.3lb psi (15kg per sq cm). A cam in contact with a roller took care of the timing in the injection pump, and control of this cam was by a centrifugal governor connected to the throttle linkage. The exact quantity of fuel supplied at a given time by the injection pump was controlled by the centrifugal governor, which in turn was controlled by the speed of the engine.

Other than the complex fuel injection system, the engine was conventional in most other respects. The single overhead camshaft was driven by Duplex chain from the forged steel crankshaft, and operated two vertically positioned valves per cylinder. All three models – 230SL, 250SL and 280SL – had sodium-filled valves to help dissipate heat more efficiently, and this is one reason why replacement valves are fairly expensive today.

Located on the right-hand side of the engine bay, the induction system was a most impressive sight, with the intake pipes protected from engine heat by an alloy shield. Quite understandably, some owners have been unable to resist polishing these components to a mirror finish, although originally all of the alloy under-bonnet components were left in their natural, dull state to minimise heat build-up in the engine compartment.

The standard four-speed gearbox with baulk-ring synchromesh was operated by a short floor-mounted lever and the gear ratios were as follows: first 4.05:1, second 2.23:1, third 1.53:1 (1.4:1 from October 1965), fourth 1:1 and reverse 3.58:1. The ratios in the four-speed automatic gearbox, which added 30lb (13.6kg) to the overall weight of the car, differed marginally and were as follows: first 3.98:1, second 2.52:1, third 1.52:1, fourth 1:1 and reverse 4.15:1. From May 1966, when the ZF five-speed manual gearbox became available as an extra-cost option, the ratios were different again: first 3.92:1, second 2.215:1, third 1.148:1, fourth 1:1, fifth 0.848:1 and reverse 3.49:1. ZF also offered a limited-slip differential as an extra-

The six-cylinder engine in all its glory, if a touch 'over-bulled', with fuel injection pump and associated pipework to the fore. The Bosch six-plunger injection pump itself (right) is an extremely sophisticated component, best likened to a miniature version of the engine it feeds.

cost option with the five-speed manual, but this is rarely found fitted to these cars today. Then as now, most owners did not need the extra traction these devices provide in difficult ground conditions, and the strange handling characteristics that limited-slip diffs can give on normal roads was another reason for their lack of popularity.

Steering was by Daimler-Benz's recirculating ball system, used in conjunction with a Stabilus hydraulic steering damper. Power assistance, an extra-cost option from the start, was a most unusual feature for a car with sporting aspirations, but entirely in keeping with the 230SL's sophisticated image and intended purpose, and much appreciated by those prepared to pay for it.

One really big improvement over the 190SL was the fitting of front disc brakes, which were made by ATE in Germany and had Girling-type calipers. Drum brakes were thought to be sufficient for the rear wheels and, apart from being made from aluminium alloy, were ribbed

similarly to the 190SL's drums to aid heat dissipation. To take the physical effort out of braking, all Pagodas had a vacuum-operated ATE brake servo, yet another non-sporting accoutrement that might have come in for criticism until those who were likely to voice discontent actually discovered its advantages.

When the 250SL was launched, it was basically the same model as the 230SL and outwardly identified only by a chromed 250SL badge on the left-hand side of the boot lid. The new 2496cc (152.3cu in) engine also looked outwardly similar to the 2.3-litre unit it replaced; although power output was unchanged, torque usefully improved to 159lb ft at 4200rpm as a result of the increased capacity. The ports were also enlarged and valve diameter increased by 2mm to 41mm (1.61in) inlets and 37mm (1.46in) exhausts to improve top-end breathing. At the same time the compression ratio was raised to 9.5:1. Although it had no effect on performance, the crankshaft

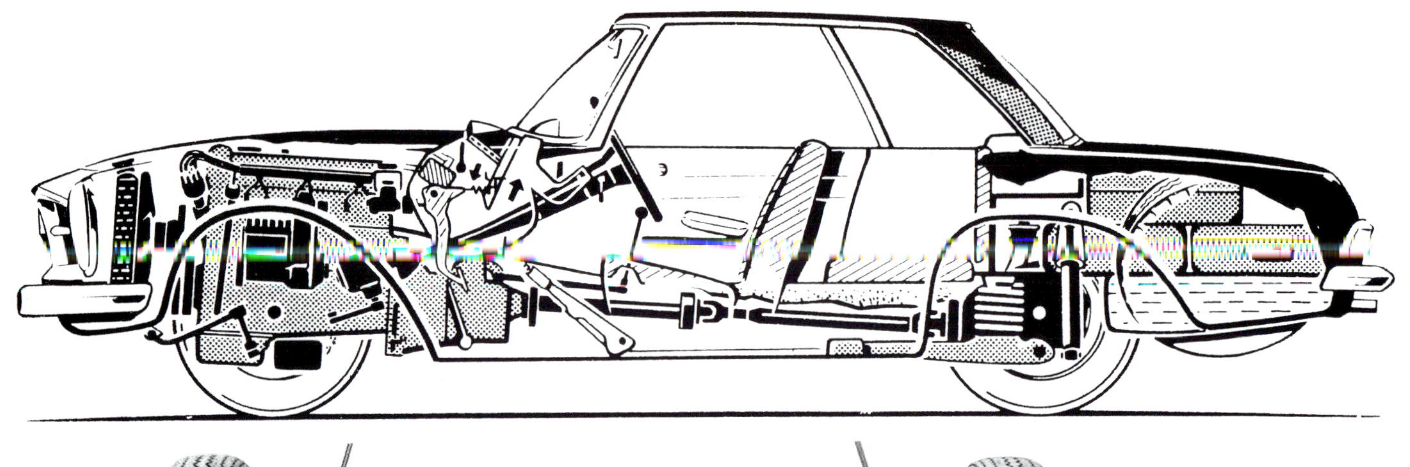

Cross-sectional sketch exposes the Pagoda's tidy packaging, while suspension views reveal the configuration of double-wishbone front and swing-axle rear; there are coil springs and telescopic dampers all round, plus an anti-roll bar at the front.

was supported in seven main bearings instead of four.

While maximum speed was a little higher than the 230SL's, by as much as 3mph (5kph) according to some claims, the great advantage of the 250SL lay in improved acceleration. And for owners who did not want to take advantage of this, the additional torque reduced the need to change gear so frequently, which obviously allowed for more relaxed driving.

The 250SL was also fitted with disc brakes at the rear instead of the 230SL's alloy drums, although many have argued that this was not strictly necessary. The handbrake operated shoes in 160mm (6.30in) diameter drums cast in the centre of the discs, which were of 279mm (10.98in) diameter. At the same time the front discs were increased in diameter from 253mm (9.96in) to 273mm (10.73in).

And then came the 280SL. This was a significantly improved car in many respects, but also a much heavier one. The engine was broadly unchanged apart from the

Various stages in development testing. The well-used S-PM 837 is seen during a towing test with a ballasted trailer and during vigorous cornering through the *Karussel* at the Nürburgring, while an early production specimen ventures on to Daimler-Benz's own test facility in Stuttgart.

obvious increase in cubic capacity to 2778cc (169.5cu in), which was achieved by increasing the cylinder bore to 86.5mm (3.41in). This change necessitated an increase in the space between the cylinders, and thus a re-cast cylinder block, a longer camshaft and reworked cylinder head. Like the 250SL engine, however, the 280SL's seven bearing crankshaft was more susceptible to the vagaries of torsional flexing than the 230SL's four-bearing item and, as a result, had to be carefully balanced with 10 counterweights hung on the throws in place of the three that had sufficed on the original Pagoda's engine. This was most definitely a case of the engineers having to take one step backwards in order to take a couple of steps forwards, but no-one – not even Daimler-Benz – had yet devised a way to make an omelette without breaking a few eggs.

Predictably, some purists scoffed at the 280SL but failed to take into account that, without the year-on year improvements and changes demanded by the marketplace, Daimler-Benz's clientele would have undoubtedly looked elsewhere to satisfy their desire for a high-protein automotive diet. Of course, the Pagoda had put on weight and, with its rubber-bushed suspension, it felt flabbier – the handling was less precise – but in this respect these vehicles reflected the aspirations of their owners to a degree, and Daimler-Benz knew considerably more about its customers than the media critics did.

The author of *Road & Track*'s August 1968 road test shed light on the changing nature of the SL breed: 'In matters of handling, brakes and ride the SL is still one of the outstanding cars of the day. Certainly its (optional but recommended) power steering is the best available anywhere, giving the impression of plain easy steering with no trickery about it. Handling is near neutral, with quick response to any steering input; but the single-pivot swing axles at the rear don't give quite the adhesion that a more up-to-date system would, so that it's easy to tweak out the Merc at the rear end. The big, sticky Firestone Phoenix radial tyres give fairly high cornering limits, but it's obvious that the added torque of the 280 engine brings the venerable suspension close to the end of its usefulness.'

Through the life of the Pagoda models, the choice available in sporting two-seaters was as considerable and wide as it ever had been, but, as with the 190SL, none could be considered a precise alternative to the Mercedes-Benz offerings. At the 'sensible' end of the spectrum Alfa Romeo offered both the 1570cc Duetto and 1779cc GTV 1750 through the 280SL period; both were

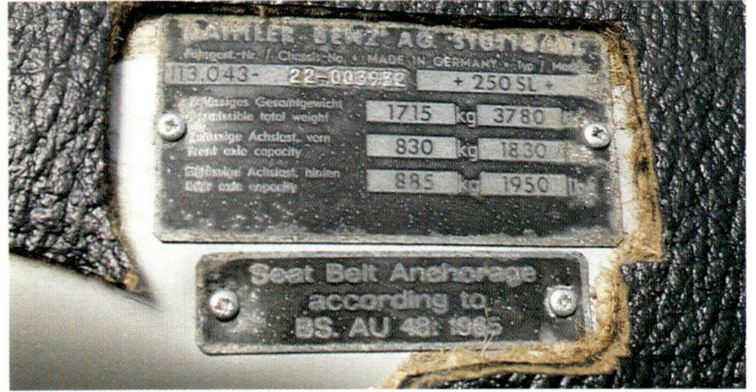

It may be externally indistinguishable from a 230SL at this angle, but this is the 250SL. The chassis plate screwed to the firewall at the rear of the engine compartment confirms the identity. When folded down into its recess, the soft-top is neatly concealed by a hinged closing panel – a beautifully engineered piece in itself.

exceptionally pretty and the GTV gave comparable performance for rather less money. Alfas, of course, were very different cars from Mercs and intended as such, but sports car buyers looked then as now at the styling first, purchase price second and performance last.

By 1968 the Austin-Healey 3000 MkIII, which had been launched at roughly the same time as the 230SL, had completely run out of steam, its prolonged development well and truly over. A 2+2, it too had put on weight and it showed, despite the availability of 150bhp from its raucous 2912cc six-cylinder engine. Even this crude, wonderful sports car – one that is still held in high regard by some senior engineering staff at Daimler-Benz – had necessarily become a shadow of its former self.

However, the most significant threat to the SL's domination of the 'Grand Luxus' sporting market in 1968 did not come from Italy or Britain, but from Germany

itself, or, to be more geographically precise, from Munich. Until this time, BMW had been involved in the serious business of making mid-range saloon cars that were reasonably quick, handled exceptionally well and sold in large numbers, but in 1968 the Bavarian company went up-market and launched a wholly new type of car in the form of the 2800CS.

This was a generously proportioned, handsome sporting coupé – a full four-seater too – with state-of-the-art suspension and 125mph (200kph) performance potential. It was no coincidence that it boasted a 2788cc (170.1cu in) in-line six-cylinder engine. Like the 280SL, there were manual and automatic versions, a high level of trim, and just as many luxury fittings. Until 1971, when it was superseded by the more powerful 130mph (210kph) 3.0CS, the 2800CS found a quite remarkable 9400 customers – and not just in Germany either.

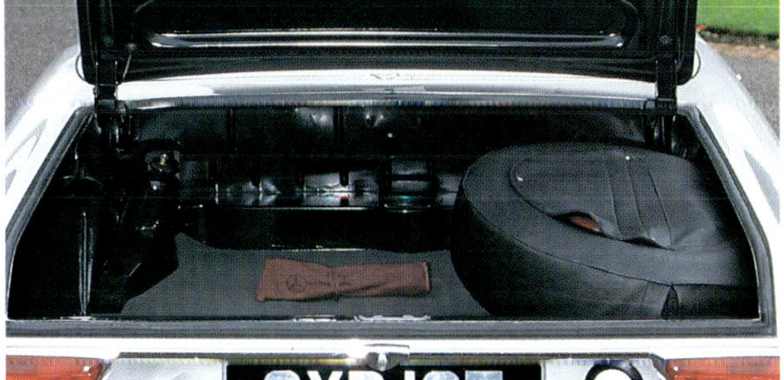

The 250SL is the rare version, made for only 11 months between December 1966 and November 1967, with 5196 built. Cars with automatic transmission had an appropriate badge fitted to the boot lid from October 1964, a change that coincided with the spare wheel being mounted flat on the boot floor rather than upright as previously.

Daimler-Benz had never taken press criticism particularly seriously, as evidenced by the broadly unchanging nature of the Pagoda range, but the threat from BMW was real and most certainly to be taken seriously. Above every other consideration, the BMW 2800CS was the impetus behind the decision to start work on the 280SL's replacement.

By the end of the 1960s the automotive world had changed dramatically – once again. Passenger safety and exhaust emissions were becoming major factors in new car design, and the 280SL felt their influence. For the North American market, the 280SL engine had to be modified to meet emissions legislation and, to this end, had revised fuel injection calibration, less valve overlap and a fuel shut-off valve operating below 1500rpm.

To counteract the power-absorbing effect of emissions equipment, Daimler-Benz took the inevitable route of developing much larger capacity engines for the third-generation SLs that replaced the Pagoda in 1971. At the same time, the trusty old swing-axle suspension was banished to the annals of history for good, and some commented that it was not before time.

During the mid-1960s many respected motoring figures, including racing drivers of the calibre of Stirling Moss, regarded the Pagoda SL as one of the finest handling, most desirable sporting cars of the day. By the

That S-PM 837 registration again, but this time the car is out on public roads after launch and not attracting much attention from the populace of Gengenbach in Baden.

end of its development, the 280SL was still a great handler and looker endowed with sufficient power to fulfil the desires of normal mortals, but the competition had caught up – and in some cases overtaken – the Mercedes in so many respects. Handling was just one area where, for example, the BMW 2800CS proved to be superior. The 280SL was still the best engineered and arguably the safest car on the planet, but it was also expensive and comparatively slow. Whereas those who understood and loved the Daimler-Benz philosophy remained ever loyal to the three-pointed star, the newly affluent, who had never owned a Mercedes, compared prices and looked towards Munich to quench their thirst for prestigious high-speed travel.

Production modifications.................

Major production changes, of course, involved the two increases in engine capacity which define the three different models in the range. However, numerous modifications were made throughout the course of the Pagoda's eight-year production life, most of them in the nature of engineering refinements. The vast majority went unnoticed by owners and journalists alike but, in true Daimler-Benz tradition, they were usually real improvements that were not made just for the sake of it.

The Stuttgart giant always employed serious engineers for whom the concept of following design whims, by making inconsequential styling changes in an attempt to increase short-term sales figures, was wholly alien. Largely due to customer and dealer feedback after the start of production, the 230SL was subjected to more modifications than the two larger engined cars, and they began to be implemented almost immediately.

The radiator was repositioned 60mm (2.4in) higher in April 1963, and the first car fitted with power-assisted steering appeared in June. The useful little 'jump' seat for the rear compartment behind the seats was offered with Tex vinyl trim from May or leather from June. A revised glovebox lid handle appeared in July, and in August the rear suspension springs were modified to reduce the vibration that they suffered on rough road surfaces.

The possibility of engine seizure was avoided by fitting new Mahle pistons with greater head clearance from September. The first car fitted with automatic transmission was completed during October, while the soft-top's fasteners were also modified that month. In November the interior door armrests were reshaped and the brake pads given a rain groove to improve stopping power in wet weather. Stronger connecting rods were employed in the engine from December.

Such was the minute attention to detail paid to these cars that the first change for 1964 involved the tool kit, which received rust-proofing surface treatment. The camshaft was tin-coated from January, while in February the horn ring was modified in shape and the steering wheel was strengthened to reduce vibration transmitted up the column. Through March, April and May a number of minor changes were made to engine components, piston clearance, for example, increasing minutely from 0.03mm to 0.05mm. Stronger door hinges, zinc-coated to protect them from corrosion, were introduced in March. From May the distributor had a modified advance curve.

Longer front springs, slightly wider 14in road wheels (6J instead of 5.5J) and the introduction of a lower steering control arm were modifications for June 1964. In

The message is one of high technology and performance – young executive and 230SL meet crew-cut pilot and Royal Canadian Air Force Lockheed Starfighter.

September a one-off car was fitted with its spare wheel mounted flat on the right-hand side of the boot instead of upright on the left, but this change, which required revised shaping to the boot floor, was not standardised until the following month. From October cars fitted with automatic transmission also received a chromed 'AUTOMATIC' badge fitted just below the 230SL designation on the boot lid. From November the soft-top was fitted with steel bows to prevent the classic ballooning effect caused by wind pressure at high speeds.

The beginning of 1965 was marked by slight modification to the front body sections to allow for the fitting of larger tyres and, as a result, the front road springs were lengthened to increase the ride height. A bezel was fitted to the glovebox lid from February but then nothing of consequence occurred until August, when the 230SL received thicker, more sharply curved seat backrests for improved comfort, new door handles, window cranks and lighting in the boot – all features that were intended for the 250SL and carried over to the revised model. At the same time American-specification cars were fitted with hazard warning lights.

Before the year was out, the exhaust system was modified and included a cast manifold from October, while the third gear ratio in the manual 'box was reduced from 1.53:1 to 1.42:1 during the same month. A larger fuel tank, up from 65 litres (14.3 Imperial gallons, 17.2 US gallons) to 82 litres (18.0 Imperial gallons, 21.7 US gallons), was introduced in November, this change necessitating corresponding modifications to the floorpan and to the longitudinal and transverse chassis members. In November the gear linkage was also modified and a new type of distributor cap was fitted.

Uncharacteristically, no changes occurred during the first ten weeks of 1966, until the floorpan was modified in March to give customers the option of fitting three-point seat belts. All left-hand drive cars received asymmetric sealed beam headlamps from April. The optional ZF five-speed manual gearbox was officially announced in May, but the transmission tunnel was not modified to accommodate it until August.

The last 230SL rolled off the Sindelfingen assembly lines on 5 January 1967 to make way for the 250SL. Comparatively few changes were made during the life of the revised model because it remained in production for only 11 months, and this 'interim' status between 230SL and 280SL makes it something of a misfit in Daimler-Benz's otherwise impeccable planning strategy. To design and produce a new engine that would see service for such a short time in the Pagoda, even if it lasted longer in other models, can hardly have been good for the nervous health of the company's financial advisers...

Besides the enlarged engine, one other significant change with the launch of the 250SL was the availability of a proper 2+2 version. A transverse rear seat had been an option for the 230SL, but now the 250SL could be ordered with a proper rear bench seat with a folding backrest, allowing two extra passengers (or additional luggage) to be carried. This feature was only available with cars ordered without a soft-top and rear closing panel. In most markets having only a hard-top for weather protection was an impractical consideration, but this 'coupé' alternative found favour in sunnier climes such as California.

Before 1966 was out, some customers caught wind of the fact that development of the new 2.8-litre engine was

A 280SL with US-only features such as distinctive headlamp treatment, side indicator repeaters and the large rubber bumper overriders – or guards – found on late cars. Tail-lamp close-up shows the amber indicator lenses that replaced red ones for all markets in February 1969. By the end of production the 280SL was heavier and its road manners less crisp, but the prestigious badge on the nose still meant everything to Daimler-Benz's loyal customers.

under way. They included Stirling Moss, who, acting against advice from Daimler-Benz, acquired the more powerful unit prematurely for installation in his new 250SL. The transplant was fraught with difficulties, as well it might have been, and shortly after the work was completed the factory officially announced the 280SL...

Official factory changes to the 250SL began in January 1967 with an alteration, confined to US-specification cars fitted with power steering and air conditioning, to the torsion bar in the bonnet counter-balancing mechanism. Special heat-absorbing glass was fitted from February, and all cars exported to North America had a 4.08:1 final drive ratio from June.

The most significant changes to the 250SL occurred in July in response to new North American safety and emissions legislation, and these were all carried over onto the 280SL. Exhaust emissions equipment was confined to US versions, but the safety alterations – designed to reduce the risk of injury to occupants in the event of a collision – were implemented on all cars.

All switches (including heating and ventilation controls) were fitted with soft plastic knobs. There was a

foamed windscreen surround, an energy-absorbing steering column, revised door armrests with softer upholstery, and modified interior door handles and window winders. A new brake fluid reservoir warning light glowed red in the event of fluid loss. There were removable windscreen latches for the hard-top and soft-top, illumination for the automatic's shift lever quadrant (a long overdue modification) and improved lighting for the heater control knobs. The Tex seat material was changed from a basketweave pattern to a coarser grain with perforations to help the seats 'breathe' more effectively during warm weather. US-specification cars also received seat belts as standard and a satin finish for the horn ring and windscreen wipers. *Road & Track*'s August 1968 test merely noted these changes and declined to comment further on them, confining criticism of the car to an ill-fitting front carpet and the passenger's window which, the writer reckoned, 'can be a pain to raise or lower from the driver's seat – it takes six and half turns'.

The last 250SL was built in November 1967 to make way for the 280SL. Apart from the badge on the boot lid, one of the few external distinguishing marks of the new

For improved safety the switchgear on late 250SLs and 280SLs was in soft plastic. The extra vents below the dashboard reveal that this US car has optional air conditioning. Fitted with complex exhaust emissions equipment, US-specification 280SLs were 15bhp (SAE) less powerful than their European counterparts.

model was the introduction of one-piece wheel trims, which were easier to manufacture than the separate hub caps and wheel trims used on the earlier models.

Because of the myriad detail changes that had occurred early in the 230SL's career, the Pagoda was rapidly running out of development potential by the time the 280SL arrived and relatively few changes were made during the production life of the big-engined model. From May 1968 all 280SLs were fitted with a safety distributor cap, and from July US-specification cars had an additional starter relay. From October US versions with sealed beam headlamps came without fog lights, and some extra drain holes were incorporated into the floorpan because of Daimler-Benz's newly-introduced system of immersing the body/chassis units in a paint bath prior to the top coats being applied. In November the brake servo was increased in diameter from 8in to 9in on left-hand drive cars, and a new power brake regulator was fitted from December.

During the 1930s, when Daimler-Benz entered the international motor racing arena, it was commonplace for teams to change final drive ratios to suit different types of circuits. Because they were so fast, tracks such as Monza in Italy and Avus in Germany demanded very high gearing, whereas Monte Carlo, for example, did not. This policy of changing rear axles continued well into the 1950s, and because Uhlenhaut was right in the centre of the action through both periods he, along with Daimler-Benz's engineering staff, appears to have become quite obsessed with gear and final drive ratios. From July 1969 US-specification cars received another change in this area when the final drive ratio was reduced from 4.08:1 to 3.92:1, and yet more alterations were to come four months later, in November, when the manual four-speed car dropped from 3.92:1 to 3.69:1 and the manual five-speed from 4.08:1 to 3.92:1.

Returning to July, cars exported to North America had changes made to the exhaust emissions control system, revisions to the sealed beam headlamps, a new ignition key switch (designed to prevent the key being removed when in position 'one'), transistorised ignition and improved fuel tank ventilation. Hazard warning lights were fitted across the range from August except, oddly enough, on cars exported to France (which had to wait until February 1970). From December the trim mouldings on the hard-top were made of aluminium alloy instead of chrome-plated brass. And finally the coolant header tank was made of plastic rather than steel from November 1970.

The last 280SL – chassis number 113 044 23885 –

The 250SL and 280SL could be ordered as a full 2+2 without a soft-top and rear closing panel. In this form, the rear bench seat could be folded to create a luggage platform.

was made on 23 February 1971. For some dyed-in-the-wool Pagoda enthusiasts it was a sad day. For others it signalled the end of a classic era and the start of an exciting new one in which sophistication, power and even greater levels of comfort would beguile third-generation SL owners for the next 18 years.

Driving impressions ....

Enthusiasts at the end of the 20th Century have become used to driving increasingly safe, fast and sophisticated motor cars. A generation ago it was inconceivable that a Volvo estate car – for so long that doyen of sensible middle-class carriages – would blossom into the 850 T5 to become a load-carrying express with a top speed in excess of 150mph (240kph). Or, to convey this startling evolution in another way, this modern Volvo is faster by some 7mph (11kph) than the Ferrari 250GT driven by Mike Parkes at Montroux on the same day that Rudi

Uhlenhaut demonstrated to the press, back in 1963, that the then newly launched 230SL could lap equally quickly.

Automotive technology has moved on at a shattering pace since those exciting days – the golden era of motoring for some – and nowadays a modern high-performance saloon can outpace just about everything made in this largely exploratory decade. However, the 230SL has to be put into context: despite the perceived vagaries of swing-axle rear suspension, which for the most part were fairly academic, most journalists were agreed that the Pagoda had exceptional road manners – superior qualities that can only be appreciated today by comparison with contemporary sporting machinery from other manufacturers.

Throughout the greater part of the 1960s, most of the SL's contemporaries were shod with crossply tyres, and many had solid rear axles suspended on the type of springs that had been used to enhance the ride comfort of horse-drawn carts and coaches during the 19th Century. Crude suspension, though, was usually forgiven by journalists – even among the most critical of the breed – if such cars were clothed in outstandingly beautiful bodies and had a powerful all-alloy V12 engine.

The Pagoda gained a certain notoriety in a James Bond movie with the baddie, Oddjob, at the wheel.

A good impression of what the Pagoda is like to drive can be gained simply by opening and closing one of its doors, then doing the same thing with a Healey 3000 or Jaguar E-type. The considerable difference encompasses almost everything for which the SL stands. Because the Mercedes was so well engineered, there is only one way to drive one properly – hard and fast! Any less and a Pagoda almost drives itself, with minimal driver input.

As the large, comfortable, body-soothing seats are fully adjustable – the ergonomic experts in Stuttgart were as capable as the engineers – finding the correct driving position takes little time and effort irrespective of your girth or height. Although many other cars of comparable quality have the supposed luxury bonus of leather upholstery, the SL's standard Tex vinyl offers better body 'grip' when side loads are imposed under hard cornering.

The controls are similarly well organised and user-friendly, as *Sports Car Graphic* noted in its May 1963 test: 'The floor-type stick shift (which is of a similar type in the automatic version) is well placed, quick and pleasant to use. The handbrake lever, however, which is located against the gearbox tunnel, has a rather crude appearance which contrasts with the general plush appearance of the rest of the car.'

This report's author, like so many of his journalistic peers, condemned the steering wheel for taking up too much room and seriously inconveniencing tall people. This aside, once you are installed behind it the luxury and quality of the cabin can still be appreciated today, as can the car's excellent all-round visibility – an outstanding achievement that brought no compromise to the car's fine looks. Using the principal functions – such as the easy-to-read instruments and the single column stalk controlling indicators, wipers, windscreen washer and lights – does little to detract from the enjoyment of driving.

Such features were also part of Béla Barényi's laudable obsession with secondary safety. The fewer distractions, he argued, the less likely it was that a driver would be involved in an accident. At the time most road testers noted this aspect, which is yet another reason why Mercedes-Benz cars acquired an unrivalled reputation as the safest ever made. By contrast so many contemporary sports cars were not, according to some, the real thing unless they were fitted with myriad switches, gauges and grotesque automotive gargoyles.

Flick the ignition key and the engine will rarely cough into life instantly from cold. This is because the fuel injection metering device does not work quite as efficiently as a conventional carburettor choke, a point readily conceded by the owner's handbook, which recommends churning on the starter until all six cylinders are firing correctly. However, when it is up and running the inherent smoothness of this beautifully balanced power unit can be readily felt even with the most fierce throttle application.

Despite its relatively small capacity, such is the fairly even spread of torque that the engine's flexibility allows any Pagoda to propel itself to top speed from as little as 10mph (16kph) in top gear, and without the characteristic transmission 'snatch' of a typically sporting, 'cammy' power unit.

In his road test for *Autosport* in 1964, Gregor Grant

The hard-top is so heavy and substantial that two strong people are required to lift it off. A minor modification occurred in December 1969 when its bright mouldings became aluminium instead of chromed brass.

enthused about the car like no other and generously praised virtually every aspect of it. His was a specially prepared press car and he wrote: 'I have been told by several people that the maximum speed of the 230SL is somewhere around 200kph or 125mph. Despite automatic drive which is supposed to affect maximum speeds, this rally-tuned machine was a great deal quicker. I pushed the speedometer needle right off the dial, and on one or two occasions saw 6400rpm on the tachometer. This is approximately 130mph and the six-cylinder engine was running like oiled silk...Acceleration was extremely good, 0-60mph taking 10.8sec, and 0-100mph 23.4sec. At 100mph cruising speeds, the exhaust note was surprisingly restrained, and there was no evidence of power roar from the induction system.'

Few failed to be impressed with either the manual or automatic gearboxes. Changing up or down with the manual is as quick as you can move your hand – it is impossible to beat the synchromesh – while changes with the automatic are almost, but not quite, imperceptible. Automatic ratios can also be selected manually and, of course, retained even at full engine revs, except when the 'kickdown' facility is used to select the next lowest ratio automatically.

Some journalists, however, found the gear ratios difficult to comprehend, as *Autocar* noted: 'Criticism of an otherwise excellent four-speed all-synchromesh gearbox is that first and third are very low. Although the engine can be taken to 6500rpm consistently, the corresponding maximum speeds are only 28 and 84mph respectively. On questioning Uhlenhaut about this, he said that a low first is necessary in the interests of good clutch life and that a high third is a peculiarly British requirement.'

Gordon Wilkins reckoned that he had never handled a more versatile or responsive automatic gearbox and, writing in *Sports Car Graphic*, was equally complimentary about other aspects of the car. It was extremely difficult even for seasoned journalists to appreciate the 230SL's dynamics, and few did until they tried for themselves some of the 'tricks' Uhlenhaut claimed the car to be capable of.

Wilkins, for example, took over the driving after some swift lappery around a racing circuit with Uhlenhaut at the wheel, and remarked: 'It seemed to be time to find out where the snags might be. I tried changing course in a corner. No problem. I shut the throttle abruptly in the middle of a full power four-wheel drift. The car slowed down but continued on course. So next I tried tramping on the brakes right in the middle of a corner while going as fast as I knew how. The car should have spun off. It held right on course. These were 100mph bends, but on the available evidence I am tempted to rate this as just

This UK-registered 280SL serves to illustrate that the hard-top could be specified in a contrasting colour – in fact there was a quite bewildering range of colours and combinations. Note also the wheel trims, of the more decorative one-piece type found on the 280SL.

about the most forgiving fast car I have ever driven. Yet it is quiet and docile with all home comforts, perfectly suitable for a dinner on a wet winter's night.'

To begin with these sentiments were echoed the world over by journalists and owners alike. And why not? The car was a sensation among the brutish, unforgiving sports cars that most were used to. By the standards of the day the 230SL's engine was under-powered – and Uhlenhaut knew it – but it is worth digesting slowly the implications of David Owen's words in *World's Fastest Sports Cars* back in 1965. He asked: 'Is this the world's fastest sports car?' And answered: 'Of course not.' But added: 'But just find one that will catch it point to point.'

This was such a perspicacious insight into what Uhlenhaut had tried to achieve – an energy-efficient, adequately-powered, crisp car that could, in the fashion of a marathon runner with the ability to sprint at the finish, quietly and confidently outpace everything else without stretching itself and getting out of breath.

All the muttering and background chirping about the still-controversial suspension system and layout simply disappeared as one journalist after another queued up to praise Uhlenhaut for taming what many incorrectly believed to be an inherently flawed design. Those who could never master the driving technique necessary for getting the best out of the Porsche 356 – and there were countless numbers who failed dismally – warmed more readily to the Mercedes because it was a much more forgiving car. It had what many regarded as the engine positioned in the 'correct' place – Porsche and Volkswagen Beetle die-hards vociferously disagreed, of course, and they had a point. The Triumph Herald – another car with swing-axle rear suspension – also had its engine in the 'correct' place, but this British saloon was a true joker in the pack because it would not handle properly, the rear

Two evocative shots, one capturing the mood, increasing affluence and freedom of the middle classes during the 1960s, the other demonstrating the versatility of the Pagoda model in all driving conditions.

wheels possessing a propensity to 'tuck in' even under braking.

By the mid-1960s little had changed in the way customers and the world's press saw the SL. Several significant modifications had been made up to this time, most of them in the nature of real improvements, but none had changed the car's impeccable manners.

And journalists continued to be impressed. In 1965 *Motor* concluded that the handling inspired a feeling of tremendous command and confidence: 'There must be a limit but on a dry road we never found it, even when accelerating faster and faster round the MIRA steering pad in second gear. Instead of breaking away the car just progressively increased its radius until tyre scrub ultimately checked further acceleration.' *Motor* discovered understeer on wet roads and tail-end break-away with fierce acceleration in the lower gears, but noted that, 'Unlike many powerful sports cars, the 230SL cannot be steered easily on the throttle'.

The optional power steering was one of the most novel features of this sporting two-seater, and possibly unique to the Mercedes at this stage in the life of production sports cars. It proved to be a great advantage over non-assisted steering because it was just as precise at high speeds but lighter and faster, which allowed

Uhlenhaut at the wheel, on the day in 1963 when he gave journalists a compelling demonstration of the 230SL's sophisticated performance. He lapped the little Montroux track in France only 0.2sec slower than racing driver Mike Parkes in a Ferrari 250GT.

mechanically sensitive drivers to place and balance the car more accurately in fast corners.

Of this *Motor* commented: 'None of our drivers have a harsh word for the power-assisted steering, which we consider to be one of the best there is. It is light, high-geared and free from kickback, though the impression of road feel that the driver gets is probably more artificial than real. The unassisted steering on the other car we tried was naturally much heavier but less precise too, and seemed lower geared. Like the clutch and gearbox, it had a rather soggy, rubbery feel and betrayed the understeer

more strongly. We much preferred the power steering.'

This last comment is interesting and indeed most surprising. At the SL's launch just two years earlier there were many who took the view that a sports car – whether it was a 'proper' one or not – should not be fitted with the kind of equipment that was traditionally reserved for flabby touring machines, but here was an esteemed journal firmly placing its neck on the block and coming out in favour of the 'soft' option.

This was Daimler-Benz right at the cutting edge. To its eternal credit, the company pushed forward new ideas – at some considerable risk to the dis-establishment of creeds that had long since been inscribed in stone – and it took many years before other respected manufacturers followed the same course. Ferrari, for example, did not introduce automatic transmission as an extra-cost option until 1976, for its 400GT model. Purists could not accept

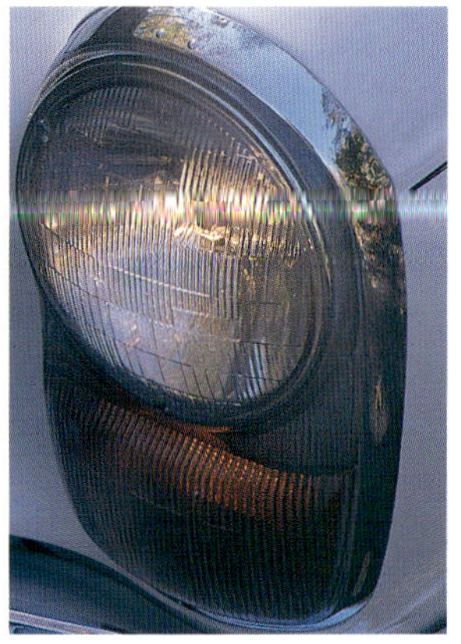

Headlamp contrast: the fully-cowled design on European models was replaced for the US by rather less elegant assemblies incorporating mandatory sealed-beam headlamps.

Transmission options contrast: the five-speed manual, available from May 1966, was not often specified, but the majority of customers chose automatic in preference to the standard four-speed manual.

this heresy, of course, but this move made sense in view of the elephantine torque produced by the 4823cc V12 engine and the laboriously heavy clutch pedal of the manual gearbox version.

The 230SL's disc/drum set-up was an equally impressive combination, but it had to be as the car's weight was a force to be reckoned with when slowing from high speeds. With the benefit of servo-assistance the pedal pressure required is minimal, and again, in complete contrast to so many contemporary 'hairy-chested' sports cars such as the Aston Martin DB4, which never needed anything less than several thigh-aching shoves on the brake pedal to bring it to heel.

Steering wheel contrast, in white and black: 230SLs and early 250SLs have a chromed horn ring flattened across the top, while late 250SLs and 280SLs have a revised horn ring that stands slightly proud of the steering wheel boss.

In *Motor's* test it was discovered that several trips through a deep water splash had no discernible effect on braking performance. The magazine's journalists recorded an increase in pedal pressure of just 7lb after no fewer than 20 half-g stops from over 70mph, and commented that such negligible fade was unlikely to be felt during normal driving conditions on public roads.

For those who were in the market for a sports car during the mid-1960s, and whose choice was based upon test figures and statistics published in the international press, the 230SL was ostensibly the least exciting among its nearest rivals. Where maximum speed was concerned it fell short of the Fiat 2300S, Aston Martin DB5, Jaguar E-type and Jensen C-V8, and the Mercedes could not outdrag any of this quartet away from traffic lights either. At 18mpg the 230SL's fuel consumption was only marginally better than that of a 4.2-litre E-type, yet the 230SL's £3866 price tag was considerably more expensive than all of these notional rivals except for the hand-built, alloy-bodied Aston Martin, which cost a hefty £4412.

However, in many respects the other cars were too compromised, and their build quality – Aston excepted – was in an inferior league. That the Mercedes was made to last is borne out by the high survival rate of Pagodas.

As is so often the case, the road test figures failed to tell even half the story, and it was just as well for Daimler-Benz that its customers chose to ignore them. The Jaguar E-type was capable of a top speed of almost 150mph, but where? By the mid-1960s traffic congestion – not to mention the gradual introduction of open-road speed limits in most European countries – was becoming such that travelling at much above 100mph was to take rather large risks with life and limb. In real-world driving conditions, where other people use the roads as well, the Merc's good, honest 115mph offered sufficient 'voltage' for most drivers.

Time and again journalists came down on the 230SL's side. Even the editor of *Motor Sport*, Bill Boddy, who rarely pulled any punches in his road tests, was hard pushed to fill the tip of his pen with poison. He found the brakes to be 'spongy' and the non-assisted steering 'tiringly heavy' at parking speeds, but he concluded: 'It is the manner in which the 230SL runs, its impeccable finish and comprehensive specification, as much as the performance which is available, that makes this a supreme sporting car for the discerning, worth the price of £3959 that it costs in this country, in the form tested.'

In December 1965 *CAR* performed an in-depth back-to-back test between the 230SL and Fiat 2300S, to some extent cars with similar backgrounds and roles. The magazine's staff judged the two cars to be very different from each other: 'Whereas the Merc lets you feel that everything is all right as long as you sit back and let the car do all the work, the Fiat demands to be driven every inch of the way...In fact the only time it really falls down is in the wet when the tail becomes skittish in typically Fiat fashion, leaving the Merc to rain (*sic*) alone.' In conclusion, *CAR* reckoned the Mercedes to be technically superior – 'a blend of sheer capability and physical comfort which has haunted sports car designers for decades' – but lacking in power and suffering from a 'ridiculously old-fashioned greasing schedule'. In most aspects the Fiat did not quite compare.

When the 250SL arrived it was much the same story, the international press remaining equally complimentary. The extra torque and all-round disc brakes were welcomed, but, as *Car and Driver* pointed out in its first test of the new car, 'The new brakes are excellent in almost every case, but the old system was already damned good'. The author of the test drove the 250SL shortly after an encounter with a Ferrari 330GTC and remarked that, 'whereas the Ferrari is designed, rightly so, for all-out hair-raising performance, surrounded by a kind of rough-hewn, brutal luxury, the 250SL is infinitely subtler. Its performance is in no way startling, yet it is capable of over-the-road averages that are limited in most cases only by the law.'

Nothing much, therefore, had changed. Some thought that the new seven-bearing engine was a little noisier, although the earlier car's power unit was also pleasantly audible under hard acceleration. The revised Mercedes still returned between 18-20mpg, and the top speed was between 116 118mph (187 190kph) depending on whether one believed the speedometer, Daimler-Benz's or journalists' figures. Some even reckoned that the 250SL was capable of more than 120mph (193kph) and said so, but, for the most part, it clearly was not.

The writer of *Motor*'s 1967 250SL road test reckoned that the extra performance was noticeable and worthwhile, 'giving this extremely handsome car the urge it deserves'. He remarked that the new optional five-speed manual gearbox, 'closed the gaps common to both the four-speed manual and automatic...the impression remains as before, that it is very difficult to go too fast for such a sound chassis...the most perfect luxury car ever to be put into production.'

The only drawback to the 250SL was that the new rubber-bushed suspension, which saved the tedious task of greasing during routine servicing, gave the handling something of a soggy feel. The car was less crisp and a little less agile. But, of course, for some the 'softening' effect of the new suspension was an advantage. *Sports Car Graphic*'s 1967 road testers pointed out that, 'The plus side of it is the fantastic lack of road shock that gets into the chassis over railroad tracks, sharp double dips and the like. Turns, lane changes, and such are accompanied by noticeable roll but, again, with no instability. In short, all

Uhlenhaut worked hard to produce a machine that could be enjoyed from behind the wheel – he so obviously succeeded!

transitions of this type, due to soft springing, seem magnified, but, unlike any other soft-sprung vehicle, the 250SL controls these moments, or couples, or vectors, so well through good geometry and substantial shocking that there are no adverse effects to either handling or stability. In fact, the control is so good that the car will take frightening dips in the road at 115mph that put all other cars we've tested up in the air.'

This test summed up the car as 'faultless' and, like so many others, its author viewed the high price tag as wholly justified. By contrast, *Autocar*'s Geoffrey Howard, writing this British magazine's first road test of the larger-engined 280SL, found that the purchase price was the only drawback to Mercedes ownership, while admitting at the same time that the car was without rival anywhere. Surprisingly, he reckoned it to be 'beautifully crisp and taut' around Brands Hatch racing circuit, but his test car was fitted with adjustable Koni shock absorbers screwed up good and tight on the rear and, as all speed merchants

are aware, a set of Konis can make all the difference to the softest suspension system...

With the extra urge of the 280SL and the molly-coddling effect of its luxurious interior, Mercedes owners soon discovered that this really was a special machine. And again most contemporary journalists failed to fault it, except that some complained – indeed as many do today about the modern breed of SLs – that it was too good, too 'clinical' and lacking in character. Geoffrey Howard wrote: 'Despite its dynamic nature, there is a kind of inanimate character about the Mercedes, nothing one can pin-point but an air of efficiency rather than spirit. It is like sliced bread compared with a crusty farm loaf.'

American journalists were also exceptionally generous in their praise of the car but, by the end of the 1960s, the majority were more at home with automatic gearboxes and Daimler-Benz's manual 'box came in for some stick. In November 1967 *Sports Car World* commented: 'The manual four is just not up to the car's high standards.

Eugen Böhringer's victory on the 1963 Spa-Sofia-Liège Rally against more powerful Austin-Healeys demonstrated the 230SL's amazing strength.

One of very few Pagoda specialists in modern-day historic rallying has been Englishman Ian Shapland, seen on a stage at Longleat in 1990.

With gear spacing giving around 30, 65, 87 and 120mph, a willowy wand and synchro lag are far less fun than one smooth surge of power with the option of flicked downshifts when the spirit moves. A better manual might help but their own kind of automatic is the real answer.'

Apart from this, there was little that anyone could say over and above what had been repeated a thousand times since the launch of the 230SL back in 1963. In theory the 280SL was a much safer car than its predecessors and, of course, a much heavier one as a result of so many running changes. But this did not matter to Mercedes owners, because it is what they had demanded. You could drive an SL across a ploughed field, with or without the hard top in place, and it would feel as solid and rattle free as the toughest 'tin-tops'. And, as with most Daimler-Benz products, the difference between the SL and 'lesser' cars is that Stuttgart's best still feels solid after many years of hard use and abuse.

By 1971, when the 280SL bowed out, the choice in the sports car car market was truly immense. Ferrari, Lamborghini, Maserati and Porsche were all producing sports cars that were blindingly fast, but financially beyond the reach of most people and costly to service.

Daimler-Benz's astonishingly loyal clientele – by this time a clearly defined breed – ignored them all. When *Road & Track* compared the 280SL with a Corvette Stingray, a Jaguar XKE (E-type) and a Porsche 911T in June 1969, the author considered the Porsche to be the best sports car among them, but summed up the 280SL with the following: 'Generalising about its qualities, it was the most comfortable, offered the least performance, has the busiest engine and the highest price tag.' He added: 'The Mercedes-Benz 280SL was a curious mixture of 'Bests' and 'Worsts'. Four out of four of our drivers

Three scenes from the 1963 Spa-Sofia-Liège show Eugen Böhringer in the 230SL negotiating a hairpin on a tarmac section (having just passed another high-tech masterpiece in the form of a Citroën DS), on a loose-surface stage (with moped rider hazardously going the other way), and in Liège celebrating victory with navigator Klaus Kaiser (Böhringer is on the right).

thought it best on ease of entrance and exit, best on driver vision, and best on heating and ventilation system. It also scored high on overall noise level and on seating comfort. But coupled to this impressive list, the 280SL scored low on engine, low on handling, low on braking and low on the appearance of the interior. Of our four drivers, one rated it as the best of the four cars tested.'

By 1970, *Autocar* had concluded that the 280SL was at last beginning to show its age compared with its rivals, but even at this late stage was moved to comment: 'Yet from behind the wheel one somehow feels that in its engineering alone it is well worth the money and at the top end of the two-seater market it has barely a single rival in the world.'

On the one hand there was a consensus that the Pagoda had become long in the tooth but, on the other, no-one disputed that Rudi Uhlenhaut had been right in his philosophy all along – which is why driving this fine car today remains the pleasure it has always been.

Motor sport

The Pagoda's competition history was fairly scant, but there was one important success for the 230SL. It is interesting that the well-known British rally driver, John Sprinzel, writing in the now defunct publication *Motor Racing*, offered a few prophetic words about the Mercedes. In an attempt to choose the best all-round rally car for international European events for the 1963 season, he commented: 'I would be inclined to favour the new Volvo, with lighter weight in two-door form, as a very strong challenger to the Mini Coopers. The Cortina Lotus seemed to go well enough on the 'Hopper', but that is a long way from proving itself over the Alpine and Dolomite passes – to say nothing of the ruts of Jugoslavia.

'From the GT side, the 2-litre Porsche is naturally an early favourite, but it will pay to watch the 230SL – should anyone take to seriously rallying one – for the suspension and roadholding will amply make up for the

Böhringer on the charge in the 1963 Spa-Sofia-Liège (above) in his battle-scarred 230SL, converted for a competitive section to 'sport trim', with two extra lamps and brightwork removed. Similar-looking car (right), equally beaten up but with a different registration, reaches the finish of the 1964 'Liège', but this time the roles were reversed with a Healey winning and Böhringer having to be content with third overall.

lack of sheer power. Sounds like an ideal car for the Liège!' Prophetic words indeed...

Daimler-Benz was no stranger to the world of rallying. The 220 and 300 saloons in particular had been successful through the 1950s and early '60s in the world's toughest events, especially in the hands of people like Eugen Böhringer. Just a few months after the 230SL was officially launched, Böhringer and his navigator, Klaus

Kaiser, were entered in a special 2.5-litre prototype for the Spa-Sofia-Liège, the world's most demanding rally.

Right from the start there was a fierce battle between Rauno Aaltonen's Austin-Healey 3000, Böhringer's 230SL and Eric Carlsson's Saab. Timo Makinen and Paddy Hopkirk put up amazing performances in their Healeys, particularly in the early stages, but heavy downpours flooded part of the route along the Adriatic and brought

Pininfarina's shapely interpretation of the 230SL poses outside the Italian design studio's Turin headquarters. This fixed-head coupé was built in 1964 as a one-off for a German businessman. Later in its life the Pininfarina 230SL went to a US owner, who changed its original silver colour for black and replaced the normal wheels with 280SL-type alloys. But the car still looks extremely handsome, its styling bearing more than a passing resemblance to contemporary Ferraris.

large boulders onto many roads, most competitors losing time as a result. Despite all of this Böhringer and Aaltonen left Yugoslavia with the loss of just four and three minutes respectively. While climbing the precarious Vivione, Aaltonen and co-driver Tony Ambrose were lucky to escape with their lives after Aaltonen misjudged a corner and terminally damaged the big Healey. This left the way open for Böhringer in the 230SL to take overall victory, with Carlsson's Saab 15min behind.

On the same event the following year, however, the tables were reversed when Böhringer finished in third place, a full 30min behind Aaltonen's victorious Austin-Healey. Sadly, Daimler-Benz then aborted its rallying efforts before making a proper return to selected World Championship events in 1978-80 – winning the arduous Safari Rallly was the big aim – with 450SLC Coupés.

In recent times, though, two British Mercedes enthusiasts have intermittently campaigned their Pagodas in a variety of historic rallies with a degree of success, adding welcome colour among the massed ranks of Lotus-Cortinas and Mini Coopers. Ian Shapland was active in national events with a 230SL in 1990-92, while Alastair Caldwell, formerly chief mechanic for the McLaren Grand Prix team, used a 280SL automatic in the 1993 London-Sydney Marathon but failed to finish owing to an axle problem.

The Pininfarina 230SL........................

Before production of the 230SL ended, the Italian coachbuilder Pininfarina produced a special one-off version for a German businessman. While the radiator grille, central three-pointed star and headlamps showed it to be nothing other than a Mercedes-Benz, the rest of the car, including the leather-upholstered interior, looked substantially different.

Bearing more than a passing resemblance to the contemporary Ferrari 250GT Berlinetta Lusso, the coupé body had steeply raked C-pillars and more rounded wheel arches, the rear wings sloped gently downwards along their trailing edges, and the front wings and transverse nose panel in front of the bonnet were lengthened. The standard tail lamp clusters were replaced by modified units from a Ferrari 275GTS and, of course, Pininfarina badges were pinned to both front wings just behind the wheelarches.

Originally finished in silver/grey, this car was later painted black and fitted with a set of the alloy wheels that were offered as extra-cost options on late 280SLs. The car, which currently resides in America, is now finished in bright red and sits on its original steel wheels. This was an interesting bespoke design exercise and one that worked well, but for the majority of owners the original Paul Bracq styling would suffice.

THE CLASSIC SLs TODAY

Despite the passage of more than 40 years since the first 190SL was made and the quarter century since the end of 280SL production, little has occurred to cool the fervour Mercedes die-hards still feel for these cars. This is not a wondrous phenomenon either, but is largely based on plain, old-fashioned common sense, for a really good original or well restored car is capable of providing exhilarating, stylish, inexpensive motoring for many years to come.

The purchase price of both the 190SL and Pagoda family remains almost incomprehensibly low by comparison with Italian near-counterparts, and restoration costs, although far from cheap, are also relatively reasonable. This is largely because SL owners have a head start on those loyal to many other motoring stables. Since the SLs were, by and large, mass-produced and sold in such large numbers, their survival rate is high and spare parts are, therefore, plentiful.

Inevitably, the 190SL is not quite so well catered for because it is so much older than the Pagodas, and locating body panels can sometimes present problems, but such is the wealth of clubs around the world that it normally only takes a telephone call or two to source even the most elusive parts. On top of this, a number of specialist engineering companies are involved in the reproduction of spare parts, and the majority work to the very highest standards. In addition Daimler-Benz's Oldtimer

A 280SL undergoing body restoration, waiting for characteristic rust in rear wheelarches and chassis legs to be dealt with.

department in Stuttgart is heavily involved in the support of classic Mercedes-Benz models of all ages and, where the Pagodas are concerned, virtually everything is available. Some bits and pieces, such as items of body trim and bumpers, can be expensive, but steel body panels tend to be reasonably priced.

Like so many cars of the 1950s, the 190SL's biggest enemy is body corrosion, simply because so few preventative measures were originally taken at the factory. No steel parts are completely immune to rust, and even the 190SL's aluminium bonnet, boot lid and doors can suffer from electrolytic reaction. The bonnet and boot lid are particularly vulnerable if the rubber seals seated between the steel bodywork and these panels perish, as they are apt to do in old age.

Although incredibly strong and durable, the box sections in the chassis frame can rot away completely if left unchecked, as can the sill panels, and when this occurs the only satisfactory course of action is major surgery. A major body rebuild with new panels requires a high degree of skill and specialist knowledge, and the cost of employing a professional to carry out such work will rarely be recouped by selling the car afterwards.

This situation also applies to the Pagodas, which is why buying and running a first- or second-generation SL should be out of enthusiasm for the cars and the SL way of life. Anyone looking to make a fast buck will almost certainly get their financial fingers burned.

As was the case with the Porsche 356, the SLs were largely ignored by classic car enthusiasts until the early 1980s – even 300SLs were relatively cheap to buy at this time – but when the market for old cars started to expand and prices began to spiral in the late 1980s a number of unscrupulous people invested lightly in large quantities of body filler, two pack paint and underseal. They 'restored' SLs to a shiny new finish, made a fortune and left the new owners of these heaps with a pile of expensive trouble on their hands.

Before purchasing any SL, therefore, it is essential to join one of the clubs and seek expert advice. Many of the 'rogue' cars, which have not been particularly well

Some corrosion trouble-spots on Pagodas are the damage-prone front cross-member (left), the top of the front wings (below left) and the base of the rear pillars on the steel hard-top (below).

restored, tend to be known to the more knowledgeable club members and their opinion is invaluable. Of course, it is possible to rebuild even the worst car, but the cost will have to be taken seriously into consideration.

Mechanically the SLs are as tough as old boots, the engines and gearboxes possessing extraordinary longevity as long as they have been maintained in a manner prescribed by Daimler-Benz. This means regular oil and filter changes at least. Failure to do this, as with any engine, will inevitably lead to rapid wear and a trail of blue smoke from the exhaust system.

A lengthy test drive will reveal if everything is in good health. Listen particularly for odd ticks from the engine's top end – the camshaft may be on the way out – and for growling main bearings at the bottom end. A 'clunking' automatic gearbox is indicative of imminent transmission overhaul, an important consideration for those on a tight budget, and problems with power–assisted steering and air conditioning will increase overhaul costs, as most of these

Matters for mechanical attention on Pagodas include corroded cylinder head waterways if anti-freeze has not been used (right), difficulty of setting up tonneau lock mechanism (below) and susceptibility of differential universal joints (below right) to break without warning.

problems need to be sorted by experienced technicians.

Generally, the really good cars are ones that are spotlessly clean inside and out, and owned and maintained by enthusiastic club members. The high prices asked for these cars usually reflect their overall condition, but before parting with cash it will still pay to have a particular car thoroughly checked by a recognised expert.

The 190SL's four-cylinder engine is engagingly simple and rebuilds are not necessarily beyond the scope of competent amateurs. Refurbishing and tuning the twin Solex carburettors, however, should be left to a Mercedes technician who knows what he is doing.

Naturally, the six-cylinder Pagoda engines are considerably more complex. The fuel injection pump alone requires precise honing, and meddlers will learn to leave it alone if they ever expect their cars to work properly again.

It is imperative to remember at all times that these cars were designed and built by qualified engineers and technicians, which is why tearing a car apart for restoration is necessarily time-consuming, irrespective of the size of your wallet. Established professionals with an international reputation for consistently producing the best restoration work – and some cars that have been rebuilt on both sides of the Atlantic are truly outstanding – are never short of work. Be prepared, therefore, to join the back of a very long queue.

Routine servicing is not beyond the scope of the average owner, of course, and the comprehensive – if rather dull looking – owner's manual gives fairly simple instructions as to how to go about it.

Despite minimal rust-proofing by the factory, the Pagodas in particular have survived the ravages of time commendably well. But the great advantage of driving a fully restored example today is that the quality of modern corrosion inhibitors makes it possible to enjoy these cars to the full all year round – although the majority seldom venture out in winter – without fearing for an outbreak of bubbling rust after every shower. Modern two-pack paints are also considerably better than the acrylic and cellulose products of yesteryear, but they contain a deadly cyanide gas and their application must be left to professionals who have the correct breathing equipment.

For those who are unfamiliar with the classic SLs, choosing the right one must boil down to personal preference and need. The 190SL is no road burner but has the ability to cruise comfortably all day at over 80mph (130kph) without stressing driver and passenger. And on the majority of roads this is quite fast enough for all except those looking for an early arrival in the next world. The 190SL has the quintessential 1950s character that is clearly lacking in the Pagoda range, its softly curving bodywork having an aesthetic charm only equalled by the 300SLs.

The 190SL is a safe, well-built car whose dynamics are best appreciated from behind the wheel. It was without such luxuries as automatic transmission and power steering, and the gearbox was always the Daimler-Benz manual four-speed, but many argue that this is a big bonus on the grounds that there is less to go wrong.

Choosing a Pagoda is a little more difficult. Purists favour the 230SL, in either manual or automatic guise, because it was the original, and the lightest, crispest handler of them all. Its four-bearing engine is also more free-revving than the later seven-bearing units. Collectors seek the 250SL because it is the rarest model. It is not appreciably faster than the 230SL, but the increase in torque offers more relaxed driving.

As it was produced in the greatest numbers, the 280SL is more plentiful and finding a good one is not difficult.

This version was better equipped and the extra power of the 2.8-litre engine inevitably has greater appeal for most. Its optional alloy wheels also enhance the car's looks.

There are no particular faults with these cars other than those that afflict all mechanical devices from time to time. The rubber gaiters on the insides of the driveshafts can split with a resultant loss of lubricating oil, but thorough checks during routine servicing are all that is needed to ensure against shortening the life of the rear axle. American-specification 280SLs fitted with exhaust emissions equipment are inevitably more complex than European SLs, but American specialists are well used to dealing with these systems.

For the reasonably affluent all these SLs remain tantalisingly affordable. Despite unrivalled engineering integrity, the prestigious three-pointed star motif in the centre of the radiator grille and perfectly adequate performance, it is unlikely that any 190SL or Pagoda will ever startle the auction world with stratospheric prices. There is a multitude of reasons for this but one significant consideration – and it is an important one – is that they lack serious competition history.

The 190SL could be converted into a 'clubbie' racer with the special equipment supplied by the factory, but successes were scarce for the few cars that were adapted in this way. The 230SL enjoyed a brief flourish as a rally car by winning the Spa-Sofia-Liège marathon in 1963 (see pages 67-70), but otherwise motor sport was a pimple on the Pagoda landscape.

The owners of 190SLs and Pagodas today do not readily identify with motor sport, the majority being content to take part in gentler club events and concours d'élégance. British and German clubs regularly organise long tours across continental Europe, giving participants a good time socially and an excellent opportunity to stretch the legs of their treasured chariots. And these cars do need to stretch their legs: Pagodas in particular need to be driven regularly, if only because long periods laid up in a garage will almost certainly lead to trouble with the fuel injection pump.

In any case, Rudi Uhlenhaut and his team worked long and hard to create a machine to be enjoyed from behind the wheel. Uhlenhaut was a driver of the highest calibre, which is why as an engineer he felt obliged to produce one of the very best cars of its day. Admiring and enjoying Paul Bracq's undated styling is fine and most do, but the real point of restoring one of these cars to its original condition today must be to experience the pleasure of what the 'old boy' was getting at all those years ago. There is nothing, absolutely nothing, to equal it.

APPENDIX

Production figures

190SL

1955	1727
1956	4032
1957	3332
1958	2722
1959	3949
1960	3977
1961	3792
1962	2246
1963	104
Total	**25881**

Chassis numbers are prefixed 121.040 (coupé) or 121.042 (roadster)

230SL

1963	1465
1964	6911
1965	6325
1966	4945
1967	185
Total	**19831**

Chassis numbers are prefixed 113.042

250SL

1966	17
1967	5177
1968	2
Total	**5196**

Chassis numbers are prefixed 113.043

280SL

1967	143
1968	6930
1969	8047
1970	7935
1971	830
Total	**23885**

Chassis numbers are prefixed 113.044

Options & accessories

During the 1950s the 190SL was considered by Daimler-Benz to be so well-equipped that the company's list of options and accessories was necessarily short. In any case a Mercedes was not the kind of car that lent itself readily to 'go-faster goodies'.

Factory extra-cost options for the 190SL, therefore, were confined to a hard-top, spot lamps, an additional passenger side wing mirror, leather suitcases, leather upholstery and a Blaupunkt or Becker push-button radio. These radios were

The tools of one of the world's classic sports combined with one of the world's classic cars: ski racks, always a popular accessory with Mercedes owners, were designed to suit both soft-top and hard-top versions of the 190SL.

fabulously reliable and always gave excellent sound quality, but unfortunately they have been rendered almost useless these days as they do not receive FM.

By the time of the 230SL's launch, Daimler-Benz had cottoned on to the increasingly large demand for additional bits and pieces. Owners with high disposable incomes were proud of their possessions, and lavished the kind of attention on them that had previously been reserved for their spouses. Most ordered a radio – usually a Becker, often by now with FM reception – but white-wall tyres, power steering, tinted glass and air conditioning were more popular in North America than other markets.

Leather upholstery, a first aid kit, vinyl covered suitcases, ski racks and roof racks, Bosch fog lamps, air horns and a fire extinguisher were all popularly sought, but few availed themselves of the underbody guard kit that was available for 'off-

High quality Becker radio sets were supplied as extra-cost options, and offered FM reception in the Pagoda era. Alloy wheels were an extra-cost option for the 280SL from 1969.

road' excursions. As these cars were popular in countries such as Africa, where some roads during the 1960s were little more than dirt tracks, it was also possible to order heavy-duty road springs to beef up the suspension.

From 1969 American-specification cars could be ordered with rather large bumper overriders, unsightly chromed items with thick rubber inserts. Far more desirable, and available from around the same time, were the lovely alloy wheels, which had 15 short spokes between the centre and the rim and roughly oval-shaped vents between them, and a cast three-pointed star in the centre. This design was appreciated immensely by Mercedes owners and a similar style was carried over to the third-generation SLs.

Of all the accessories that were available, the most essential today is a fire extinguisher. Although these cars were no more prone to setting themselves alight than any other vehicle from this era, the carrying of a fire extinguisher is usually a condition of obtaining classic car insurance nowadays.

Body colours...................................

190SL

Regular body colours

Black (DB40), White (DB50), White Grey (DB158), Blue Grey (DB166), Silver Grey Metallic (DB180), Graphite Grey (DB190), Pearl Green (DB213), Light Blue (DB334), Delphin Blue (DB380), Fire Engine Red (DB534), Strawberry Metallic (DB543), Ivory (DB608).

Special order body colours

Pearl Grey (DB125), Grey Beige (DB157), Dark Grey (DB164), Light Stone Grey (DB169), Cement Grey (DB186), Lime Green (DB218), Dark Green (DB221), Moss Green (DB226), Mid Green (DB229), Blue Green (DB270), Light Green Metallic (DB274), Mercedes Blue (DB335), Mid Blue (DB350), Light Blue Metallic (DB353), Light Blue (DB356), Brazil Brown (DB409), Beige (DB412), Dark Tobacco Brown (DB423), Cream (DB439), Light Beige (DB441), Mid Brown (DB442), Mid Red (DB516), Red (DB519), Red (DB538), Ivory (DB620), Cream (DB629), Yellow (DB630).

Coupé/hard-top combinations

Body colour	Hard-top
Black (DB40)	White (DB50)
	Silver Grey Metallic (DB180)
White (DB50)	Black (DB40)
	White (DB50)
	Light Blue (DB334)
	Fire Engine Red (DB534)
	Graphite Grey (DB190)
Blue Grey (DB166)	White Grey (DB158)
	Black (DB40)
White Grey (DB158)	Graphite Grey (DB190)
Silver Grey Metallic (DB180)	Graphite Grey (DB190)
Graphite Grey (DB190)	Black (DB40)
	White Grey (DB158)
	Silver Grey Metallic (DB180)
Pearl Green (DB213)	Black (DB40)
Light Blue (DB334)	White Grey (DB158)
Delphin Blue (DB380)	White Grey (DB158)
Fire Engine Red (DB534)	Black (DB40)
	Ivory (DB608)
Strawberry Metallic (DB543)	Silver Grey Metallic (DB180)
Ivory (DB608)	Black (DB40)

230/250/280SL

Body colour	Code
Black (230/250/280SL)	040
White (230/250/280SL)	050
Arab Grey (230/250/280SL)	124
Light Grey (230SL)	140
White Grey (230/250/280SL)	158
Blue Grey (230/250SL)	162
Anthracite Metallic (230/250/280SL)	172
Anthracite (280SL)	173
Mid Grey Metallic (230/250SL)	178
Silver Grey Metallic (230/250/280SL)	180
Light Beige (230/250/280SL)	181
Graphite (230/250SL)	190
Moss Green (230SL)	226
Dark Green (230/250/280SL)	268
Dark Olive (250/280SL)	291
Horizon Blue (230/250/280SL)	304
Dark Blue (230/250SL)	332

Light Blue (230/250SL)	334
Blue (230/250SL)	335
Mid Blue (230/250/280SL)	350
Blue Metallic (230/250/280SL)	387
Mid Blue Metallic (230/250/280SL)	396
Havana Brown (230/250SL)	408
Tobacco Brown (280SL)	423
Dark Red Brown (230/250/280SL)	460
Bronze-Brown Metallic (230/250/280SL)	461
Tunis Beige Metallic (230/250/280SL)	462
Copper Metallic (230/250SL)	463
Sand Beige Metallic (280SL)	467
Red (250SL)	501
Mid Red (230SL)	516
Light Red (230/250SL)	519
Dark Red (250/280SL)	542
Lasu Red Metallic (230/250SL)	567
Signal Red (230/250/280SL)	568
Red Metallic (230/250/280SL)	571
Dark Bordeaux Red (230SL)	573
Red (280SL)	576
Light Ivory (230/250/280SL)	670
Grey Beige (230/250SL)	716
Beige Grey (280SL)	726
Beige Grey Metallic (280SL)	728
Papyrus White (230/250/280SL)	717
Moss Green Metallic (230/250/280SL)	834
Blue (280SL)	903
Dark Blue (280SL)	904
Light Blue Metallic (280SL)	906

Soft-top colour	**Code**
White Grey (230/250SL)	132
Brown (230/250SL)	437
Cream (230/250SL)	702
Light Grey (230/250SL)	708
Black (280SL)	740
Dark Grey (280SL)	741
Light Grey (280SL)	742
Parchment (280SL)	743
Dark Blue (280SL)	744
Beige (280SL)	745
Dark Brown (280SL)	746
Dark Green (280SL)	747
Brown Beige (230/250SL)	863
Black (230/250SL)	872
Marine Blue (230/250SL)	896
Dark Green (230/250SL)	970
Green (230/250SL)	972
Beige (230/250SL)	979

Technical specifications ....

190SL

Engine In-line four-cylinder **Construction** Alloy cylinder head, cast iron block with alloy crankcase/sump unit **Crankshaft** Three main bearings **Bore x stroke** 85.0mm × 83.5mm (3.35in × 3.29in) **Capacity** 1897cc (115.7cu in) **Valves** Single overhead camshaft driven by Duplex chain from crankshaft, operating two vertically positioned valves per cylinder **Compression ratio** 8.5:1 **Fuel system** Bosch mechanically driven fuel pump, twin Solex 44PHH horizontal (or sidedraught) carburettors **Maximum power** 105bhp (DIN) at 5700rpm **Maximum torque** 104.9lb ft at 3200rpm **Transmission** Four-speed manual with full synchromesh, 200mm Fichtel & Sachs clutch **Gear ratios** First 3.41:1, second 2:1, third 1.29:1, fourth 1:1, reverse 3.2:1

Final drive ratio 3.7:1 **Top gear mph per 1000rpm** 18.2mph (29.3kph) **Brakes** ATE hydraulic system with cast iron Al-fin style drums **Front suspension** Unequal-length wishbones, coil springs, telescopic shock absorbers, anti-roll bar **Rear suspension** Swinging half axles, trailing arms, coil springs, telescopic shock absorbers **Steering** Recirculating ball with Pitman arm, unequal-length tie rods, hydraulic damper **Wheels/tyres** 5.0×13 five-stud steel wheels, 6.40-13 Continental crossply tyres **Maximum speed** 105mph (169kph) **0-60mph** 11.3sec **Maximum speed in gears** First 32mph (51kph), second 55mph (89kph), third 85mph (137kph) **Length** 169.9in (4290mm) **Width** 68.5in (1740mm) **Height (unladen)** 52.0in (1320mm) **Wheelbase** 94.5in (2400mm) **Front track** 56.3in (1430mm) **Rear track** 57.9in (1470mm) **Turning circle** 36ft (11m) **Ground clearance** 6.1in (155mm) **Kerb weight** 2510lb (1140kg) **Permissible total weight** 3080lb (1400kg) **Payload** 570lb (260kg) **Maximum front axle load** 1500lb (680kg) **Maximum rear axle load** 1580lb (720kg)

230SL

Engine In-line six-cylinder **Construction** Alloy cylinder head, cast iron cylinder block, alloy sump pan **Crankshaft** Four main bearings **Bore x stroke** 82.0mm × 72.8mm (3.23in x 2.81in) **Capacity** 2306cc (140.7cu in) **Valves** Single overhead camshaft operating two vertically positioned sodium-filled valves per cylinder **Compression ratio** 9.3:1 **Fuel system** Fuel injection with six-plunger injection pump **Maximum power** 150bhp (DIN) at 5500rpm **Maximum torque** 144.5lb ft at 4200rpm **Transmission** Four-speed manual (optional five-speed from 1966) or four-speed automatic, 200mm Fichtel & Sachs clutch **Gear ratios** Four-speed manual: first 4.05:1, second 2.23:1, third 1.53:1, fourth 1:1, reverse 3.58:1. Four-speed automatic: first 3.98:1, second 2.52:1, third 1.52:1, fourth 1:1, reverse 4.15:1. Five-speed manual: first 3.92:1, second 2.215:1, third 1.418:1, fourth 1:1, fifth 0.848:1, reverse 3.49:1 **Final drive ratio** 3.75:1 (4.08:1 with five-speed) **Brakes** ATE dual circuit hydraulic with front discs and rear drums **Front suspension** Double wishbones, coil springs, telescopic shock absorbers, anti-roll bar **Rear suspension** Low-pivot swing axles, coil springs, telescopic shock absorbers **Steering** Recirculating ball, hydraulic steering damper, optional power assistance, 3.2 turns lock to lock **Wheels/tyres** Steel 5.5×14 five-stud steel wheels, 185HR-14 Continental or Firestone Phoenix radial tyres **Maximum speed** 115mph (185kph) **0-60mph** 10.5sec (11.5sec auto) **Length** 168.7in (4285mm) **Width** 69.3in (1760mm) **Height (unladen)** Coupé 51.4in (1305mm), roadster 52.0in (1320mm) **Wheelbase** 94.5in (2400mm) **Front track** 58.0in (1474mm) **Rear track** 58.5in (1487mm) **Turning circle** 33.6ft (10.35m) **Kerb weight** 2855lb (1295kg) **Permissible total weight** 3638lb (1650kg) **Permissible front axle load** 1764lb (800kg) **Permissible rear axle load** 1874lb (850kg)

250SL

As 230SL, except: **Crankshaft** Seven main bearings **Bore x stroke** 82.0mm × 78.8mm (3.23in x 3.10in) **Capacity** 2496cc (152.3cu in) **Compression ratio** 9.5:1 **Maximum power** 150bhp (DIN) at 5500rpm **Maximum torque** 159lb ft at 4200rpm **Brakes** ATE discs all round **Maximum speed** 117mph (188kph) **Front track** 58.4in (1484mm) **Rear track** 58.5in (1485mm) **Kerb weight** 2998lb (1360kg) **Permissible total weight** 3781lb (1715kg) **Permissible front axle load** 1830lb (830kg) **Permissible rear axle load** 1951lb (885kg)

280SL

As 250SL, except: **Bore x stroke** 86.5mm × 78.8mm (3.41in × 3.10in) **Capacity** 2778cc (169.5cu in) **Maximum power** 170bhp (DIN) at 5700rpm **Maximum torque** 177lb ft at 4500rpm **Maximum speed** 121mph (195kph) **0-60mph** 9.3sec **Wheels/tyres** Alloy wheels optional from 1969

ACKNOWLEDGEMENTS

Grateful thanks are due to the owners whose cars feature in special colour photography by Simon Clay (UK), James Mann (US) and Dieter Rebmann (Germany): they are Brian Parker of Star Quality (US 190SL), John Williams (UK 190SL), Joachim Stickel of Stickel Oldtimer-Restauration (German 230SL), Mike Bailey (UK 250SL) and Nick Soprano of Motor Classic and Competition Corporation (US 280SL). Many of the black and white photographs came from the Mercedes-Benz Classic Archiv in Stuttgart, where Fred Langer provided special help. Other sources of illustration were Nick Kisch, Neill Bruce (including the Peter Roberts Collection), David Hodges, Quadrant Picture Library, *Classic & Sports Car* magazine, Dennis Adler, Hans Wiborg-Jenssen and Ian Shapland.